Agile Changed My Life

How to Manage the Chaos of Life and Career with Agility

D. Ray Freeman

D. Ray Freeman

Copyright © 2014 D. Ray Freeman. All rights reserved.
Published by Agile Consultant Guide, LLC

No part of this publication may be reproduced, stored in a retrieval system or transmitted in any form or by any means. Electronic, mechanical, photocopying, recording, scanning or otherwise, except as permitted under Section 107 or 108 of the United States Copyright Act, without either prior written permission of the Publisher, or authorization through payment of the appropriate per-copy fee to Agile Consultant Guide, LLC. Requests to the Publisher for permission should be addressed to the Contracts Department, Agile Consultant Guide, LLC, 8551 Boat Club Road Suite 121-167, Fort Worth, TX 76179, (800) 536-6130 or online at www.agileconsultantguide.com.

Limit of Liability/Disclaimer Warranty: While the publisher and author have used their best efforts in preparing this book, they make no representation or warranties with respect to the accuracy or completeness of the content of this book and specifically disclaim any implied warranties of merchantability or fitness for a particular purpose. No warranty may be created or extended by sales representatives or written sales materials. The advice and strategies contained herein may not be suitable for your situation. You should consult with a professional where appropriate. Neither the publisher nor author shall be liable for any loss of profit or any other commercial damages, including but not limited to special, incidental, consequential or other damages.

For general information on our other products and services or for other technical support, please visit us on the web at www.agileconsultantguide.com or email us at info@agileconsultantguide.com.

For all Corporate and Government Sales call (800) 536-6130 or email us at profiles@acg1.net

ISBN: 069202803X
ISBN-13: 978-0-692-02803-2
Library of Congress Control Number: 2014914501
AGILE CONSULTANT GUIDE, FORT WORTH, TEX.

To my loving wife, Tahonie; my mother, Sonja; my sister, Karen; and all of my aunts, uncles, cousins and friends who have contributed so much to my life thus far: thank you. I dedicate this to you as well as to our loved ones who are no longer with us: my father, Donald Ray Freeman Sr.; my brother, Howard W. Pope; my aunts Shirley J. Depillars and Abigail C. Spencer; and my dear cousin Brittany Smith.

D. Ray Freeman

Sometimes it snows in April.
Sometimes I feel so bad.
Sometimes I wish that life was never ending.
But all good things, they say, never last.

—Prince, "Sometimes It Snows in April"

Howard's greatest wish was to find peace of mind. A talented musician and decorated military veteran, he ended his battle with depression on July 4, 2013. My brother Howard was a smart and generous person, willing to help anyone, do anything, anytime. By updating my understanding of the brain, I have gained a new appreciation for it and a new respect for his struggles. It is my goal to carry on my brothers' generosity by using my talents and ambitions to help people improve their lives and careers.

Depression is the leading cause of suicide and sadly, it affects more than forty-thousand Americans every year. Many of us do not understand that depression is a mental disorder. It manifests in the brain and causes uncharacteristic behavior. As with any disorder, there are treatments that can help but, there is much need for improvement. Luckily, there are people around the world using their talents and ambitions to help improve the treatment of mental disorders. One day, they will find a cure.

In tribute to my brother, Howard W. Pope, a portion of the proceeds from the sale of this book will directly benefit the Brain & Behavior Research Foundation. The foundation's efforts are helping to improve treatments, find cures, and reduce the devastating effects of mental illness on so many people and their loved ones all over the world. The Brain & Behavior Research Foundation is committed to alleviating the suffering caused by mental illness by awarding grants that will lead to advances and breakthroughs in scientific research and development. For over a quarter of a century, the foundation has awarded more than three hundred million dollars worldwide to more than thirty-three hundred scientists carefully selected by its prestigious Scientific Council. And 100 percent of all donor contributions go directly to the researchers, thanks to the foundation's operating costs being generously underwritten by two family foundations. Throughout its short history, the foundation has become the go-to organization for researching disorders in children and adults such as depression, bipolar disorder, schizophrenia, autism, ADHD, PTSD, and anxiety.

I encourage you to make a donation today at:

www.bbrfoundation.org/donate

Our family and friends are extremely grateful for your generous support.

D. Ray Freeman

Contents

    Dedication   vi

    Acknowledgements   xi

1   Introduction to Agile@Home
Life Is a Project Worth Managing   13

2   What the Heck Is Agile?
Getting Familiar with Agile from Values to Principles   27

3   Building Your Career: Freelancing for Companies
Success Tools for Independent Agile Consultants and Contractors   55

4   Agile Frameworks for Achieving Your Personal and Professional Goals
Scrumming Your Way to Success in Everything You Do   79

5   Learning over Education
The DALA Principle: Developing a Learning Aptitude toward Your Goals   113

6   From Now On, I Will Plan My Day in an Agile Way
Realizing Your Dreams One Chunk at a Time   137

7   The Working Agreements of Love
Change Your Routine, Change Your Environment, Change Your Life   153

8	**If You Had More _____, What Would You Do with It?**	
	Commitments, Rewards and a Unique Use of Agile	165
9	**In Closing**	175
10	**Resources Recommended by the Author**	179
	About the Author 185	

Acknowledgments

I would like to thank my beautiful wife Tahonie for all of your support in turning a dream into reality. Thank you for allowing me to share a part of what makes our lives special with the rest of the world. Thank you to Dr. Geil Browning of Emergenetics® International for sharing your wisdom in such an impactful way. Thank you to Jason Waite Photography for making my big head look presentable on the cover! Special thanks to Ed Grannan of Improving Enterprises and the members of AMP for your inspiring input. Thank you to Jeff McGuire of Fortis Talent for placing me in the right place at the right time. To Elaine McNaughton and Ethan Hodges for your coaching and inspiration. To Gerry Robert and Les Brown, thanks for turning a seminar talk into a life-changing experience. Special thanks to Kimberly Smithson-Abel, Jean Paul Goconnier, Kelly Tisdell, John Jacobs, Jonathan Ball, Virginia Farris, Lily Johnson, and Pete Chambers for your roles in introducing me to Agile..

"Align yourself with people that you can learn from, people who want more out of life, people who are stretching and searching and seeking some higher ground in life."

—Les Brown

Chapter 1

Introduction to Agile@Home

Life Is a Project Worth Managing

Success is not final, failure is not fatal: it is the courage to continue that counts.

—Winston Churchill

D. Ray Freeman

mypokcik/shutterstock.com

If your life were a project, what kind of project would it be? Would it be a discovery initiative, a constant search for meaning, intellect, and purpose? Or one of development, a relentless pursuit of growth and progress? Perhaps you'd prefer a problem-solving mission, one that seeks to reduce life's sufferings.

In fact, your life is composed of several projects, all running concurrently, all competing for your attention. Some are urgent, others can wait. Many of these projects are things on your to-do list that have just accumulated over time, but some of them are special. Yeah, some of them are dreams, goals, and aspirations way beyond the honey-do list or the mama-do list. They may be things you've moved to the back burner, eventually taking them off the stove completely.

Your career is a project. Your job, your business or whatever you do for money or happiness are all parts of your career project. And just like life, these little projects are all running at the same time, competing for your attention. Some are urgent, others can wait. You understand how important it is to enjoy your work. And you want to enjoy it. Whether you value your work because it's really easy or extremely challenging, the bottom line is that you want more. Well, guess what? You deserve more. And, you can have more. But, there is a catch. You have to be willing to do something about it. Dreaming of better days is a good start but eventually, you have to dissect those dreams into smaller, workable parts.

Corporations do a great job at this because they have processes in place for people to dream or brainstorm regularly. They come up with ideas and strategies that will help them achieve their goals. They have groups of people dedicated to their success. They try new things. It may not always work out perfectly but they have strong motivations to be better than average. They want more. They believe in the fact that they deserve more are willing to do something about it.

And it's really as simple as that. We all have different projects going on in our personal lives and in our careers. We should all dream and brainstorm on a regular basis, coming up with our own ideas and strategies for success. We should try new things and even if they don't work out perfectly we should have strong motivations to be better than average. We should all be able to say that "I want more and I believe in the fact that I deserve it. And most importantly of all, I am willing to do something about it."

Multitasking is the simultaneous management of two or more tasks by an individual. Although it is incredibly inefficient, most of us find ourselves doing it anyway. We do this because we have so many little projects, with so many little tasks going at the same time. It's hard to manage all of this chaos!

A project is a collaboration of efforts, planned and designed to achieve a goal. This definition is full of keywords that must find a permanent place in your vocabulary. When competing projects are left unmanaged, a chaotic environment results. Chaos is a state of utter confusion and disorder. If your life or career has ever been in a state of chaos, then you know the pain and the discontent that comes from this type of environment.

Not many of us have a proven way to manage our projects. This is a fundamental mistake many people make. Just like a business, you need to find a way to manage your chaos. And since every individual and every family is unique, your chaos and mine are not the same. However, the underlying methodology used to manage each one of our respective projects can be the same. We can all be Agile.

Agile offers a human-centered approach to managing the chaos of life and career interdependently. Technically, Agile is a project management method, but it has evolved to form a new generation of leaders who boost the performance and profitability of companies. Agile has frameworks that can be customized to align the endless flow of priorities accumulating in the backlog of your life, emphasizing those that will deliver the most benefit toward your goals.

An Agile approach is one that creates a productive environment that can remain that way indefinitely. In the workplace, Agile offers tremendous benefits to employers and employees. When adopted by business leaders, it speeds up their response to market change. Executives can manage their portfolios strategically, teams can self-organize with a sense of autonomy, and self-actualization finds its roots. Agile is evolving the modern business landscape. By applying its basic values and principles to everyday life, you can put your life and career on the fast track to success.

What all can you do with Agile? Like the executives, can you get a higher rate of return on your investments too? Time is an investment, and so is effort. Both are as valuable as money, and you deserve to get a return on the time and effort that you invest. Agile is a great way to manage your projects in a way that achieves the highest return on investment (ROI) on all of your life's investments.

Corporations value time and effort and recognize both as investments from which they expect a return. Unfortunately, most people place little or no value in their time or effort, but all of life's little projects are hungry for both. If you don't feed them they will surely starve to death.

So what if you could initiate a process that would speed your ideas to a place where they can become reality? I bet you could apply a greater focus on the pursuit of the life you want to live, the person you want to be, and the impact you want to have on the world. As you incorporate the tenets of Agile in your own projects, you'll develop a mind-set that consistently creates opportunities. Opportunities for you, your family and loved ones.

Being Agile can change your life in the most positive ways imaginable. I want to share with you some of the ways Agile changed my life. You'll discover a much better method for managing the chaos of life and career with agility—where methods you may have used on a project in your work environment can bring substantial benefit to the management of projects in your home and personal life.

The features you want represented in your life may be stuck inside of some other big idea that's not moving. But what good are ideas if you never take action on them? Have you ever said to yourself, "I want to do something different but I'm not sure what?" A little prioritizing can help you figure that out. The Agile mind-set is resourceful and takes ownership of learning and development on an ongoing basis. As you learn new information, you'll do a better job of prioritizing what to do first, what comes next, and what won't get done at all.

A goal is the intended result that effort is directed toward. Effort is the energy that drives you toward your goals. Many of our goals become blocked because excuses get in the way. Think of someone you know who always has an excuse to justify their misfortune. You'll probably hear them say things like, "I can't do this because I don't have that". Or, "I can't do something because I don't have what they have." Some people have been blocked for so long that it feels normal to them. Don't become one of these people. Release your blocks, don't embrace them. Removing those blocks will unleash the true power of effort.

It's easy to get stuck on something and find yourself doing the same things over and over, expecting a different result that may never come. With Agile, if you get stuck on something, you'll recognize it sooner rather than later and look for ways to get unstuck. With every endeavor, you'll learn to adapt and adjust as needed. You'll find yourself evaluating whether something is "done" or "*done*-done" and create a common definition that distinguishes one from the other. By becoming more Agile, you'll be able to implement features into your character that will empower you to drive change as well as adapt to the unexpected—everything from changes in your work life and personal life to changes in your mental and physical well-being.

If life is the totality of its features in various stages of completion, then we live in an iterative cycle of works in progress. We have a growing number of interrelated projects in our backlogs, each with a unique range of dependencies, just as within a complex organization.

Throughout the creation of this book, I tried to imagine life through the lens of the information technology world I've lived in for almost two decades. In this view, life is a portfolio made up of several programs, each with a set of projects ripe with dependencies, tasks, risks, rewards, and competing objectives. I can hear the language of stories being written to express my purpose and the goals I wish to achieve in life—goals that are pertinent to the development of my character, my abilities, and the impact I have on others. I would eventually discover that Agile methodology could be useful for doing just about anything I could put my mind to.

One of the goals of this book is to promote the benefits of being Agile within and beyond the workplace. For those who are new to all of this, I'd like to welcome you to world of Agile and give you the language and tools for successful Agile adoption. You'll be able to take an Agile approach to anything that's important to you. If you're hoping to get through personal hardships, make more money, lose weight, or just find a better way to keep track of your kids' household chores, being Agile can change your life.

For those of you who are already Agile savvy, you will find some unique ways to be Agile in other areas of your life. You'll recognize how Agile can be used to drive your personal motivation for success. Agile provides a familiar framework to apply the wisdom we learn from reading books and attending classes—As well as what we learn in retrospect of our most recent experiences. This book may even be that first step to introducing Agile to your significant other.

Agile is well-aligned with many of the values endorsed by top motivational speakers, coaches and trainers. Setting goals, planning, being open to change, striving for excellence, self-organizing and reflection to name a few. In fact, Agile is in alignment with the way our brains are genetically wired to learn new things.

The word neuroplasticity refers to the brains ability to change and adapt in response to learning. With a little repetition, you can train your brain to direct your thoughts and actions along the path of your choosing. The more you focus and practice doing something the better you become at it. Studies in modern neuroscience show that by doing this, new neural connections are created in the brain. These new connections allow you to develop new skills, overcome obstacles or form new habits.

Imagine if you could train yourself to be more successful or teach yourself a new set of skills that will make you healthier, wealthier and happier. Agile is the ideal framework for you to initiate this mental stimulus.

Agile techniques encourage such insightfulness. Writing down your goals, posting them in a place where you can see them, talking about them and working on them with other people all inspire your brain to take action. Repeating these actions that will help you form a recurring habit of seeking out success.

Imagine if this becomes second nature to you or a regular activity in your household, as it is in mine. You can teach yourself to grow emotionally and prosper financially. As you do this, people will take notice of your new growth, but they probably wouldn't understand it. Some might call it luck or say that you are blessed beyond your worthiness. But, I believe that applying Agile techniques in your personal life can help you become a better person and definitely a better communicator.

Our abilities to be successful are directly influenced by how well we communicate. Having good communication skills can boost your capabilities in the workplace and at home. Agile simplifies communication to a level we can all understand—The kindergarten level! Some of the things we value in the Agile workplace are behaviors most of us learned in our first few years of school. For instance, we express ourselves. We talk, laugh, read stories and play with stickers during class. We adorn the walls with pictures that show our imagination and our progress. We play little games that entertain and educate us at the same time. As we age, most people discard that childlike vigor in favor of more sophisticated behavior. But elementary acts of collaboration are far more effective than people give them credit for. These simple activities are just as valuable to your mental development today as they were back in kindergarten.

I will show you some of the tools that I use to help people understand themselves better. And, how you can establish a common understanding with other people about how you communicate best. You may even discover some hidden strengths that have been blocked for so long, you never knew they existed before. If you are feeling stuck in any kind of way, you should write yourself a note right now that says: **"I will find a way."**

You'll learn to develop a learning aptitude for the most important things in your life and use your own strengths to accelerate your pursuit. I'll share some experiences that will offer insight into ways that Agile can change your life.

I finally know what I want to be when I grow up. I want to be someone who helps people succeed. I welcome this as an opportunity to transcend the environments where I first found Agile. If there was one statement to define the vision I have for the readers of this book, it would be the following:

> As an author, I want to inspire you to do everything you do better—not just in your career, but literally everything. By merging superior mental awareness strategies with an Agile approach to life, you will find ways to create a resourceful, self-driven environment that is beneficial to creating and engaging opportunities and navigating any course.

I can honestly say that being Agile has helped me to become a better person, a better team player, and definitely a better leader. I am confident that it can do the same for you. I will share the highs and lows of my experiences as transparently as possible, artistically giving you the details of how *Agile Changed My Life.*

Chapter 2

What the Heck is Agile?

Getting Familiar with Agile from Values to Principles

The principles we must universally accept are the recognition of our shared humanity, our shared aspiration to happiness, and the avoidance of suffering. From these principles, we can learn to appreciate the inextricable connections between our own well-being and that of others.

Together, they constitute an adequate basis for establishing ethical awareness and the cultivation of inner values. It is through such values that we gain a sense of connection with others and it is by moving beyond narrow self-interests, that we find meaning, purpose, and satisfaction in life.

—The Dalai Lama, *Beyond Religion*

iQoncept/shutterstock.com

Agile in the Software World

Agility is defined as the ability to move quickly and in a coordinated manner. An agile person is mentally acute and aware of self, situation, and environment. This is a fitting title for the software development methodology that bears this name. Being Agile is being efficient and effective; not being Agile is being lethargic and awkward. Speed and efficiency are the fundamentals of successful software development, as they allow companies to gain a competitive edge over their competitors. To keep up with rapid change in the technology world, companies establish procedures that can be repeated consistently.

In the past, all software development was approached sequentially with a set of procedures called "waterfall." Large chunks of time were dedicated to defining the solution and capturing the intricacies of each requirement. Development, testing, and deployment were phased approaches. Each stage depended on the completion of its predecessor before it could start. All work was performed according to a strict set of requirements that were gathered, defined, and elaborated over several consecutive months.

For decades, this approach provided an effective way to build systems by first writing a full set of upfront specifications, but it was not so effective in providing incremental value. Nobody got a chance to see or benefit from any working pieces early on in the process. Change is inevitable. However, change is disruptive to the waterfall process; therefore, it was usually discouraged and unwelcome. A modification to scope would have to be traced and managed to avoid causing rework and confusion for those downstream.

The waterfall approach required more people, more time and was slower to deliver its return on investment. Although Agile-based methods are growing rapidly, waterfall still exist at many companies today. Part of the maturation of Agile teams is being able to coexist with other methods seamlessly.

Waterfall Methodology: the predecessor to Agile, waterfall breaks a project into stages:
- Requirements gathering
- Design
- Coding
- Testing

In a Waterfall process, each step must be completed before moving on to the next, and all steps in the process must be completed before any value is delivered to the customer. The name "waterfall" is comes from development process that literally flows from one stage to the next.

Agile Methodology: "Agile" is the umbrella term for iterative, incremental software development methodologies, including:
- Extreme Programming (XP)
- Scrum
- Crystal
- Dynamic Systems Development Method (DSDM)
- Lean
- Feature-Driven Development (FDD)

Agile methodologies arose in opposition to the traditional waterfall methodology. For the purposes of this book, we will focus on Scrum, which is the most widely used today. Agile emphasizes small teams delivering small increments of working software, with great frequency, while working in close collaboration with the customer and adapting to changing requirements.

Where Did Agile Come From?

In 2001, at a ski resort in Utah, seventeen independent thinkers came together to address the need for a non-documentation driven approach to software development. Kent Beck, Mike Beedle, Arie van Bennekum, Alistair Cockburn, Ward Cunningham, Martin Fowler, James Grenning, Jim Highsmith, Andrew Hunt, Ron Jeffries, Jon Kern, Brian Marick, Robert C. Martin, Steve Mellor, Ken Schwaber, Jeff Sutherland and Dave Thomas were all present. Together, they devised a whole new approach to software development, one based on a set of core values and principles. It would be based on people instead of bureaucracy and would help answer the real questions of what people want and what can people deliver.

Founder Robert Martin made the resounding statement, "At the core, I believe Agile methodologists are really about the 'mushy' stuff—about delivering good products to customers by operating in an environment that does more than talk about 'people as our most important asset' but actually 'acts' as if people were the most important, and lose the word 'asset.'"

The Agile movement was not readily adopted in the business world. It was seen as the rogue anti-methodology, shunned for its perceived lack of planning, documentation, and structure. But in fact, it was just the opposite. Agile evolved into a state of being, grounded by a philosophy that transcends software development and has benefits far beyond the workplace. The Agile Manifesto, available at www.agilemanifesto.org, was penned by the original founders on February 13, 2001. Since then, Agile has grown into many different frameworks, disciplines, and approaches, all based on this human-centric philosophy.

My Introduction to Agile

It's a beautiful Friday afternoon here in the great state of Texas. The sun is shining bright, but I wouldn't know it because I'm sitting at my desk, surrounded by gray cubicle walls. Like most consultants, my workspace is decorated the same way it was the day I got there: bare, with only the basic essentials to do my job. Karen, my boss, walks in.

With a loud thud, Karen drops two books on the desk in front of me. "I need you to do something," she says with eyebrows raised.

In typical fashion, I open with a joke. "What can Brown do for you?"

We both have a good laugh, and she replies, "You can deliver!"

One of the books is titled *Getting Real: The Smarter, Faster, Easier Way to Build a Successful Web Application*, by 37signals. The other is *Agile Estimating and Planning*, by Mike Cohn. "Ray, I need you to read these by next week. We've got a coach coming by to give us a hand."

"A hand with what?" I reply with a confused expression.

"We're going to bring in a new way to build software that'll be faster and better than before—oh, and you're gonna implement it."

As a former waterfall project manager, I was first introduced to Agile in 2004. What began as an approach to software development eventually morphed into other uses of Agile inside and outside the office. I discovered that the same values and principles could be applied in my approach to career and life achievements in general. I began to apply elements of Agile into my own life. In other words, I became Agile.

Many practitioners of Agile have come to realize and accept this viewpoint. Being Agile has cultured and organized many areas of my professional and personal life. The benefits have been profound. Since becoming Agile, I've been able to refine and release new features of life more effectively. and I believe the same can happen for you. In fact, Agile can strengthen anyone's ability to accomplish anything. People who work in Agile environments are equipped with a powerful set of tools to aid in the pursuit of success and happiness. Leaving these tools at work at the end of the day is an injustice to you and those who could benefit from a better you.

> **The Agile Manifesto**
> The Agile Manifesto is a set of values and principles adopted to uncover better ways of developing software by doing it and helping others do it.
> www.AgileManifesto.org

What Does It Mean to Be Agile?

Agile is composed of four values and twelve principles. Applying one or more of them, by definition, is being Agile. Doing more is more Agile. The approach's original intention was to uncover better ways to develop software, but over time, Agile itself matured. It's a process that makes sense and can help you make sense of things that don't, especially when adapted for everyday situations you encounter.

Agile Value 1

- We Value **Individuals and Interactions** over Processes and Tools.

Agile professionals work closely with one another, usually in the same room and interacting with each other on a regular basis. This can be uncomfortable for "techies."

Most technical skills are expressed in front of a screen or book or when engaging in some other solitary activity. To comply with the first value of Agile, you have to get in touch with the social side of your brain.

It boils down to this: if you can communicate with people in a face-to-face conversation and establish a common understanding, you gain a lifelong skill that is beneficial to everything you do. Make an effort to understand and appreciate the intricacies of human interaction. Agile is more than a process or a set of tools; it's a way of thinking. When people work together in this way, they often produce long-lasting friendships, something we could use more of in every community around the world.

Everything is about people. Everything you do, the way you do it, and whether you do it at all will have an effect on you as well as others.

Agile Value 2

• We Value **Working Software** over Comprehensive Documentation.

The waterfall process follows a meticulous plan that produces a lot of documentation up front. These documents become a static guide for building and implementing the final solution. Depending on the size of the effort, this can last for months. The problem with this is that during those months, no value is produced. This may be OK if you're producing something that has very little likelihood of changing or when change is impractical. Those in charge of manufacturing and construction projects often prefer this because after-the-fact changes can be costly.

Software projects are more fluid, and the abilities to changing and adapting quickly are endearing characteristics of Agile. Modern software products are made up of several individual units called modules. These modules are self-contained units of working software, each adding its own layer of increasing value as it is released.

Time is valuable and irreplaceable. As soon as a usable piece is complete, it should be put to use right away. Businesses know this, but most people don't realize the cost or the value of time. The sooner that some working module can be put to use, the sooner you can start capturing value from it.

As a personal value, you are most effective when you can deliver solutions to life's challenges in a modular fashion. A single chunk of work may not appear to have much value on its own, but as you assemble the pieces, you'll begin to see your ambitions take shape in a whole new way.

When you build your life modularly, every piece of every feature is useful and beneficial on its own. Instead of taking months or even years to map the perfect plan, focus on the small things, those that allow you to taste success one sip at a time. **Done is better than perfect**.

Agile Value 3
- We Value **Customer Collaboration** over Contract Negotiation.

The customer is the end user of whatever it is you're in the process of creating. In this case, your customer may be anyone that you serve or produce a result for, including your loved ones or even yourself. In Agile, collaboration occurs among the decision makers (the businesspeople), solution makers (the technical people), and customers (the end user) throughout the entire engagement. Since working software is delivered in small chunks, collaboration with all parties is essential to make sure those chunks represent part of the vision that you're trying to make into reality.

In more rigid environments, all of the requirements for the complete solution are negotiated up front. Changes are not encouraged, are usually sealed under contract, and are modified only through a strict change process. The customer must sign off on a completed set of requirements, and the team builds the solution according to the agreed-upon design parameters.

By adopting this value, you agree to work together with the people who have a vested interest in your endeavors. What's yours becomes ours, and collectively we can accomplish anything that we put our minds and efforts to.

Agile Value 4

• We Value **Responding to Change** over Following a Plan.

This value is very clear. It's what makes Agile so applicable to each and every one of us. Have you ever made a long-term plan for your life and followed it step-by-step without making any changes? How'd that work out for you? I'm guessing not so well. There are lots of people out there who are focused on an outdated plan. Change is one thing in life that is always certain. Instead of responding to change, people tend to wait for it to pass so they can hop back on the old plan.

The technology world provides a good example of this. The hottest new thing today will be obsolete by next quarter. It troubles me to see people passionately pursuing things that require a high up-front cost, produce nothing for large amounts of time, and have little potential for the type of reward they're looking for. From a corporate standpoint, this is best explained through the story of Company Zero.

The Story of Company Zero

Company Zero sells all of its products on the web. As an online retailer, it has seen continuous growth over the past three years. However, a new mobile technology has been released that makes it easier for customers to buy products like Company Zero's via a mobile device. The website is very successful, so the company invests a lot of money on projects to enhance it. Company Zero has plenty of projects in the works to make it easier for customers to buy more online but nothing in the works for mobile devices.

This quarter, Company Zero's online sales dropped by more than 25 percent. Its customers have been complaining that they can't buy stuff with their phones, so they're leaving and going somewhere where they can.

Company Zero is in trouble. It's losing money today and investing in the wrong thing for tomorrow. This could be the end of Company Zero. It has already invested too much to abandon its existing projects, which are all somewhere between the design and implementation stages. If the company were to cut one of them now, hundreds of thousands of dollars would be wasted. All it would have to show for the effort would be a stack of project documentation and fragments of useless code that may never be implemented.

But what if Company Zero remains faithful to its original plan and continues with its current portfolio of dot-com enhancements? Eventually, the company will successfully release each one of them, victoriously claiming it has completed the work to the specifications of the requirements.

Value 4 allows an Agile team to switch lanes quickly without losing the momentum already established. Business leaders can navigate the workforce to where it's needed with little or no loss. Agile can navigate the course of business, which protects it and you, from the effects of financial loss.

If Company Zero keeps hemorrhaging money, it will have no choice but to start cutting jobs.

And so it came to be. Company Zero closed its doors for the final time. Hundreds of people found themselves unemployed, not due to lack of ability, but due to lack of agility.

> At regular intervals, Agilists reflect on ways to become better at what they do. They see change and make the necessary adjustments to move forward. This provides an opportunity to delve deep into what's working well, what's not working well, and what could use some improvement. Adjusting your path early is key to avoid making costly mistakes in your business and personal life.

The Twelve Principles of Agile

Alongside its four core values, Agile is grounded by twelve principles. Each of them is a tenant of software philosophy, although easily relatable to the endeavors of life that we all pursue.

1. **Our highest priority is to satisfy the customer through early and continuous delivery of valuable software.**
    - Satisfy your needs and the needs of others by delivering value to yourself and to them, early and often.
2. **Welcome changing requirements, even late in development.**
    - Life is full of changes. Welcome and respond to change, no matter when or how it arrives.
3. **Deliver working software frequently, from a couple of weeks to a couple of months.**
    - At regular intervals, deliver something of value—something that works. Do this habitually and frequently. Done is better than perfect.
4. **Business people and developers must work together daily throughout the project.**
    - In everything you do, be sure to collaborate with those around you and strengthen your ability to work together. The efforts of many outweigh the efforts of one.
5. **Build projects around motivated individuals. Give them the environment and support they need, and trust them to get the job done.**
    - Surround yourself with motivated people. Create an environment of support and trust

that will enable you and those around you to be successful.

6. **The most efficient and effective method of conveying within a team is through face-to-face conversation.**
    o Share information in the most effective manner possible, preferably face-to-face. Understanding how people convey and perceive information is paramount to building an environment of good communication.

7. **Working software is the primary measure of progress.**
    o Working results, no matter how small, are the primary measure of progress and the building blocks to success in any situation.

8. **The sponsors, developers, and users should be able to maintain a constant, sustainable pace indefinitely.**
    o Agile processes should be realistically sustainable and able to maintain a constant pace indefinitely. Don't work yourself to death.

9. **Continuous attention to technical excellence and good design enhances agility.**
    Always be improving your life and the lives of those around you. In addition to this, you should always be improving your environment and

building on a foundation that allows you to scale in the direction you're headed.

10. **Simplicity—the art of maximizing the amount of work not done—is essential.**
    - There's no need to overcomplicate things for the sake of completion. When the goal has been achieved, the work is done. Once your appetite has been satisfied, stop eating. The goal is to realize the vision, not to build all of the pieces.

11. **The best architectures, requirements, and designs emerge from self-organizing teams.**
    - Autonomy is the ability to govern and direct your own actions. When people feel free and are engaged in their work and with the people they work with, they don't require any prodding to carry out their responsibilities. Happy people are the most productive people in the world.

12. **At regular intervals, the team reflects on how to become more effective and then fine-tunes and adjusts its behavior accordingly.**
    - This is one of the most important things a person can do. Stop, look back on how things went, and ask yourself, "If I could do that over again, knowing what I know now, what would

I do differently?" It's best to ask yourself this question every couple of weeks rather than every couple of years. Never dwell on the past but always have a retrospective of what went well, what didn't go so well and what you would change.

> **Scrum in a Nutshell:**
> - Split the work toward your goals into a list of small, concrete deliverables placed on sticky notes on a board. Sort the list by priority and estimate the effort of each item.
>
> - Split time into short iterations or sprints (usually one to four weeks), with stuff you can finish and put to use by the end of the sprint.
>
> - Review what you did at the end of each iteration and update priorities and plans.
>
> - Keep refining the process by having a retrospective and action items to take after each iteration.

Frameworks for Agile Teams

Agile can be practiced as an overarching philosophy that guides the path of businesses. It's effective from the executive levels down to the day-to-day operations. At the team level, it's usually practiced by groups of ten or fewer.

In the software world, Agile comes in several disciplines, which are also called frameworks. Each Agile framework follows the same values and principles but from a slightly different approach. For the purpose of the illustrations described in this book, we will focus on just one of them—Scrum.

In Scrum Agile there are three main roles: the scrum master, the product owner, and the team.

Scrum Master

The scrum master serves as the team's coach. He or she provides the leadership and motivation to propel the team toward its goals. A good scrum master uses the values and principles of Agile to help team members work together in the most effective way possible. The scrum master serves the needs of the team by removing roadblocks that impede its progress.

The title of scrum master is usually used synonymously with project manager, but there are some distinctions between the two. Since a majority of Agile is based on communication, part of the scrum master's job is to facilitate open discussion between people. In my experience, the best way to do this is to establish a culture of cognitive diversity across teams.

Workplace diversity has many forms. These include distinctions such as age, gender, ethnicity, religious beliefs, political views, sexual orientation, heritage, education, disability, and life experience. Cognitive diversity is the tolerance, understanding, and inclusion of what's going on in each of our brains.

Good scrum masters are in touch with their team on a cognitive level. Beyond keeping track of progress and dealing with risks and issues, the scrum master is responsible for radiating Agile philosophy throughout the team.

Product Owner

The product owner is responsible for the business aspects of a project, such as ensuring that the right product, service, or result is being built in the right order. A good product owner can balance competing priorities, is available to the team, and is empowered to make decisions about the product, service, or solution that is being developed. He or she represents the voice, the eyes, and the ears of the customer.

The Team

The team itself is made up of the individuals who play whatever roles are necessary to complete the project or goal. Good teams self-organize and assume whatever roles are necessary. In the software world, these include business analysts, architects, developers, testers, and visual designers, just to name a few. But to be Agile outside the office, the team can be members of your family, friends, or those you interact or conduct business with. Although the scrum master provides guidance, it is the team members who collaboratively decide which person should work on what tasks. They also define ways to maintain increasing levels of quality, efficiency, and effectiveness.

Key Elements of Scrum

- ➢ Time-boxed iterations, usually one to four weeks in duration
- ➢ The team commits to a specific amount of work for each iteration
- ➢ Uses velocity as the metric for planning and process improvement
- ➢ Cross-functional teams
- ➢ Items must be broken down so they can be completed within one iteration (sprint)

- Amount of work in progress is limited per iteration
- Work item size estimation is required
- Cannot add items to an ongoing iteration
- Prescribes three roles (product owner/scrum master/team)
- A scrum board is reset between each sprint
- Requires a prioritized product backlog

To benefit from Agile independently, you may find yourself playing any or possibly a combination of every role on your projects. Some projects are very personal but have a lot of requirements, tasks, risks, and dependencies that still need to be dealt with. You will undoubtedly find situations where you are the architect, the developer, the tester and the customer. You think it, you build it, you make sure it works and you buy it.

For instance, as the product owner of your life, you have the ability to write the stories that make up the backlog of what you will and will not do. As the scrum master, you are responsible for removing obstacles and making sure the team can perform. Where the product owner wears the business hat firmly, the scrum master must embrace the softer side of humanity, which thrives on inspiration and ability. You'll have to get to know yourself and those around you well enough to seek and provide inspiration where needed.

I want to make it clear that I'm not proposing that life become rigid. There are things in life that are made better by the experience of not knowing. Being Agile in life is not the removal of spontaneity. One of the things you will love about it is the fact that you can continuously create fertile ground for new opportunities to evolve naturally.

Generically, Agile is easy to learn, implement, and train others on. Regardless of what knowledge you currently have or will ever acquire, becoming Agile will improve the way you set and reach goals as well as identify and solve problems.

Regardless of job title, industry, or experience level, you can develop and nurture the success you seek. Regardless of age, culture, or gender, you can achieve any goal by incorporating Agile values and principles creatively throughout your life.

Lessons from Mom

My mother taught me to make a conscious effort to help others and to be the type of person others could emulate. In honor of her, I have embarked upon a mission to help people understand themselves better and build a better understanding of the people around them. Along this journey, I have come to understand myself better, as well as the people around me. I am grateful for her inspiration to be of assistance to others and to use and refine strengths to overcome weaknesses, all while recognizing that I am a work in progress, capable of changing lives, both mine and yours.

Chapter 3

Building Your Career: Working for Companies You Don't Work For

Success Tools for Independent Agile Consultants and Contractors

When it can be said by any country in the world, my poor are happy, neither ignorance nor distress is to be found among them, my jails are empty of prisoners, my streets of beggars, the aged are not in want, the taxes are not oppressive, the rational world is my friend because I am the friend of happiness.

When these things can be said, then that country may boast its constitution and government. Independence is my happiness, the world is my country, and my religion is to do good.

—Thomas Paine

D. Ray Freeman

Kheng Guan Toh/shutterstock.com

Job Independence—Contractor or Consultant

I was originally introduced to the world of job independence after being laid off. If you have had this experience before then you can definitely relate to this statement: When your income suddenly stops, you'll do whatever it takes to make it start back up again. The first job offer I received was a temporary assignment managing a mid-sized system implementation. The pay wasn't great, the location wasn't ideal but I needed work and was in no position to be picky. When I got the job offer letter in the mail, I shared it with my grandmother. In her generation, temporary work was frowned upon. You get a job and stick to it for life! I can almost hear her voice saying to me: "Baby, maybe if you do a good job they'll keep you around longer."

In the early 2000s, it seemed that even the tech industry itself was confused about the distinction between consultants and contractors. The industry has since matured, and luckily, I have too. Working as a project manager can take your career into many different industries and translates to many different job titles. Of the many directions I could have steered my career toward, I remained in e-commerce development, as well as with Agile.

An ever-growing number of companies are promoting the use of Agile throughout all of their technology initiatives. They need people who are Agile to help others become Agile. Agile helps to foster some of the critical success factors that determine the effectiveness of any team, communication and planning being the highest. Company CIOs know this well and are focused on upholding this environment in their organizations. It's quite a challenge for them to make sure these success factors exist among employees, but it is even more challenging when a part of the contributing workforce is composed of people who work there but are not employed there.

Although they may hold the exact same position in a company, there is a slight difference between contractors and consultants. They both provide an important benefit to their clients, but one approaches their work differently from the other. Consultants and contractors both play a unique and critical role in the success of the companies they serve.

Contractors in the technology industry are usually ex-employees of other companies that had to trim "headcount" and are looking for a permanent work solution sooner or later. This can be a great strategy as it gives you the opportunity to earn a significant income while you and the company learn more about each other. Both sides get to determine if they are a good fit for the other. Adventurous types gain the opportunity to check out multiple environments before settling down to one.

The Contractor Approach

➢ Contractors are extensions of the employee base.
- o A good contractor will blend in with employees and become a good fit for a future permanent position with the company.

➢ Contractors usually want permanent jobs.
- o Contractors are concerned with not having the securities usually associated with permanent employment. Health benefits, company perks, retirement plans, profit sharing, training, and personal development opportunities are normally not offered to contractors. They are willing to temporarily forgo these luxuries in exchange for a higher pay rate than the average employee in the same position.

➢ Contractors follow the rules, whereas a consultant is more likely to break them.
- o A mentor of mine once told me that "contractors do what they're told, and consultants tell you what to do." The contractor's job is to be an

extension of current staff, working under the direction of their assigned manager.

The Consultant Approach

➢ Consultants possess some degree of partiality to being independent.
- A consultant is usually not looking for a job. He or she is either a business owner trying to grow the firm's personal offerings or is employed by a larger consulting company. Those who are employed this way still favor independence as their assignments are meant to be temporary in nature.

➢ Consultants are hired to do what's necessary, not to do what they're told.
- When a company brings in a consultant, it's usually with the understanding that he or she has a particular expertise that may be lacking internally. Their role is to drive what needs to be done in order to overcome challenges.

➢ Consultants bring in new ideas.

True consultants will shake things up if necessary. Their value is heightened by their ability to challenge the status quo of companies and bring about change in an innovative way.

Depending on your circumstances, it may be beneficial to employ a *consultant's approach* to the way you perform your job. There are not many guides on moving from contractor to consultant and quite often, both words are used synonymously. If you decide to make the change to an independent career, it is important to distinguish yourself as more than someone just filling in temporarily. No matter how long you are assigned there, it is your responsibility as a consultant to add a unique value to your clients work environment.

Being Agile helped me make the transition from employee to contractor to consultant. Agile is a great process to assist you in guiding your career path. People often tell me that they want to become more than a contractor but also want more flexibility than an employee. The transition will require some extra efforts on your part. You'll have to claim your independence and own it.

Have you made the decision to declare your job independence? Before you make this important decision, be aware that as an independent worker, you probably won't be provided with everything you need to be successful. The security of full-time employment will tempting. Although the perks and benefits of traditional employment have their appeals, they cast a small shadow on the virtues of freedom and self-reliance experienced by job independence. What the contractor may view as lack, the consultant sees a wealth of opportunity.

As mentioned before, the role of the contractor is to augment the existing workforce on a temporary basis or until a permanent job opening becomes available. Many of the people who work as contractors today are products of an earlier downturn in the technology industry. The dot com bubble that burst in the early 2000s created mass unemployment across the country. Many knowledgeable employees were out of a job and competing for full-time work. Companies struggled as well and some were unable to add "headcount" to their employee base. However, they could continue their growth efforts by adding temporary workers. These temporary workers were really seeking to replace their previous jobs but were willing to accept a contract-to-permanent position as an interim solution.

Freelancing Agility

The independent Agile consultant could be any person who plays a role on a development team including the hierarchy of managers. As the economy breaks through ceilings and crashes through floors with certain ambiguity, technology sectors rely on an increasing amount of temporary workers to supplement their workforce.

According to the Eighth Annual State of Agile Survey, conducted by software maker VersionOne, In 2010, 7.4 percent of all workers were contractors. By 2013, that number grew to 12 percent. The increase of these working arrangements generate substantial economic and other values for both workers and employers.

This allows companies of all sizes to use labor services in situations where a traditional employment relationship is either impractical or uneconomic.

Independent consulting is widespread in industries where workers move from project to project frequently or work on multiple projects at once. Companies need to be able to respond to short-term changes in demand. There is often a need to fill supply gaps by calling on more workers than they could economically maintain as traditional employees.

This presents a unique challenge to most companies. It can be difficult to integrate temporary team members into existing team environments. Part of that difficulty is getting them up to speed, adapted to the corporate culture, and able to add value as fast as possible. Temporary workers are often the least tenured people in a department. It's tough for a company to instill a sense of ongoing camaraderie when some of those people have no desire to permanently join the company.

The independent workforce in IT are like technical mercenaries: hired guns to get in, get the job done, and get out. As an independent consultant, you are expected to walk in the door already equipped with the knowledge, tools, and aptitude required to meet the needs of your clients. Those who can do this are also able to increase their bill rate and become highly sought after in their field.

Independent workers have a great deal of freedom in deciding where, when and how to pursue their careers. But keep in mind that the companies you work for share this freedom as well. Temporary workers should always be prepared to find new assignments and adapt to new environments whenever the need arises. To put it bluntly, you can become jobless rather quickly. So keep your skills sharp, your senses aware and your finances in order. Know your needs and how to take care of them.

Maslow's Hierarchy of Needs

Maslow's hierarchy of needs is a theory in psychology proposed by Abraham Maslow in 1943. Usually represented by a pyramid as shown in the following illustration, this hierarchy remains a stable framework of modern management training, education and human motivation. It applies to every worker—employees, consultants, and contractors alike. The independent worker must be aware of his or her needs and implement ways to achieve them in terms of their career as well as their personal endeavors.

SELF-ACTUALIZATION
morality, creativity, spontaneity, acceptance, experience purpose, meaning and inner potential

SELF-ESTEEM
confidence, achievement, respect of others, the need to be a unique individual

LOVE AND BELONGING
friendship, family, intimacy, sense of connection

SAFETY AND SECURITY
health, employment, property, family and social abilty

PHYSIOLOGICAL NEEDS
breathing, food, water, shelter, clothing, sleep

Elenarts/shutterstock.com

Employee versus Consultant

Most companies are heavily invested in meeting the needs of their employees. They provide an environment that does a decent job at serving a majority of these necessities. However, there are usually some gaps in the self-esteem and self-actualization departments. Satisfying these desires can be challenging for businesses. Outside of traditional human resources efforts, it's hard to justify the cost or time investment to achieve such elusive results.

Even if such services are offered by a company, you may not have access to them as a consultant or contractor. In fact, you will probably be expected to have these skills as a prerequisite to your employment, and rightfully so. Regardless of your role, when you are independent, you are expected to be a leader, self-manager, and self-organizer. It is ultimately your responsibility as a professional to build and develop your competency as a mature individual.

As an Agile consultant, you should be able to translate these abilities to others in a way that grows and develops them as well. It's an awesome environment when people refine their skills independently because it leads to team interdependence.

Workers choose an independent career for any number of reasons. Some are averse to joining an organization because they don't wish to play by the rules. They don't like the idea of being limited to a traditional job situation. Some have a bit of history that influences their decision of whether to "marry" the company or "just stay friends." The distinction of liberty can provide a new outlook on your career options and the ability to increase your income substantially. It provides a level of satisfaction that can actually increase your effectiveness in your profession.

> **Interdependence**: Relying on mutual assistance, support, cooperation, or interaction among constituent parts or members

Agile is a tool of my profession that I also use to grow my career. The results of this combination have been life changing in my response to adversities as well as accomplishments. It is my sincere hope that you will seize the opportunity to experience the same results. You deserve it.

DBAC: Doing Business as an Agile Consultant

Businesses make money; they make progress. Some of them are pretty good at it. Those who are successful probably got that way because they do something well. They put processes in place to help them grow and do more things well and well things better. Ultimately, they strive to be the best. This is the same level of intensity that should resonate from you not only as a consultant but also as a person. Put a process in place to help you do more things well and well things better, ultimately striving to be your best.

Successful people and successful businesses have a lot in common. Both are driven by a passion-fueled vision—never dwelling on the past.

> If you are a consultant, you are a business. If you're going to do business, you'll need to act like a business. Otherwise, get a job!

They both put forward a considerable amount of effort to understanding where they are today and how to get where they're going tomorrow.

Businesses have people who work together toward common goals that achieve business value. Consultants should conduct their businesses in a similar fashion. If you want to do what successful people do, you'll have to do what successful people do!

Recruiters and Recruiting Relationships

Building your business as an independent consultant doesn't mean that you have to do it alone. To establish and maintain your career, you will have to build a strong team around you. Establishing a relationship with a good recruiter is essential to career independence. I emphasize *good recruiter* because it is important to recognize the difference between good recruiters and bad recruiters. Both will pull your résumé if its posted on an online job board, especially if it contains keywords that match a job requisition they are trying to fill.

> Independent Agilists help companies deal with some of their toughest challenges, professionally and with superior return on investment. A good Agile consultant can enhance corporate culture, inspire leadership, and stimulate change throughout an organization.

Make sure your résumé has the exact words that match the type of job you want. Some recruiters pull résumés from websites in bulk using digital algorithms. These algorithms automatically search for specific words and rank them by the number of times they appear in the document. This is only a bad practice if it is the only differentiator in whether they contact you for an interview or not.

Good Recruiters versus Bad Recruiters

Good and bad recruiters are both capable of finding work for you; however, the good ones do a better job at it. When you publish your résumé on one of the job boards such as Monster.com, Dice.com, or Careerbuilder.com, the first people that will usually respond are recruiters. They don't work for the companies that you're trying to work for, but they are acting as agents for them. A good recruiter has a personal relationship with the hiring managers of their client companies. They get to know them and understand their needs as well as their business culture.

The strength of this relationship is one of the differentiators between good and bad recruiters. When a recruiter has a collaborative relationship with his or her clients, he or she is in a better position to bring in the appropriate people to fill roles.

Bad recruiters lack this relationship. The worst ones go on a fishing expedition. They scour the Internet for company job openings with little or no knowledge about the client or the company. They get a list of keywords plucked from job descriptions and go on the hunt. There's rarely a shortage of people looking for work, so candidate résumés are easy for them to find. They give false hopes to job seekers by responding with positions that have little relevance to their preferences or capabilities. A bad recruiter can actually hurt your career because he or she will submit you to anything that's open, hoping to make a commission. They prey on people's anxiety of being out of work and desperately try to squeeze square pegs into round holes. Hiring managers suffer as well, because they have to weed through dozens of unsolicited résumés that may not fit their needs.

Good recruiters are more than just headhunters. They have a deep understanding of the work environment. They make it a point to find the "right fit" for their clients and won't approach them with anything less. They usually have a few key accounts that they're primarily focused on, and they will take the necessary time to get to know you before presenting your résumé to their clients.

Good recruiters, also referred to as vendors, are usually willing to help you to grow, whether you are working through them as a contractor, filling in for staff, or as a consultant, resolving a more specific need for their clients.

> In 2001, I was the project manager on a team implementing a customer relationship management system. It attached to an Oracle database on the backend. So the word "Oracle" appears *once* on my résumé—from thirteen years ago. Occasionally, I still get a call from a bad recruiter, offering me the perfect Oracle database administrator job!

The Most Valuable Skills Are Not on the Résumé

To gain a recruiters point of view, I spoke with Benica Brown, CEO of Connections IT Services. This Dallas-based company is small but has developed close business relationships with several Fortune 500 clients. Their high level of communication and personalized support provided to consultants is remarkable. This dedication gives Connections IT Services the ability to compete and win in the highly competitive staffing industry.

In a 2014 interview, I asked Mrs. Brown to elaborate on the topic of training and career development for independent workers. "Competency development should be a collaboration of the vendor, the consultant/contractor, and the client company," she said. "For short-term engagements, the responsibility lies solely with the individual. The client company needs to provide some type of training in the specific way it does business to get the person up to speed quickly in his or her environment."

We spoke about the role of money in the process of charting your career path. "Sometimes it's worth it for the consultant or contractor to select a role that pays less but offers more opportunity in the future," she explained. "You may have to take a step back in order to take several steps forward. The most successful consultants are those who are willing to make a conscious effort in creating their future."

She went on to say, "Personality and attitude outweigh skill set. Consultants should always conduct themselves in a manner that promotes a sense of global leadership. Where technical skills can be taught, personality and attitude must be developed. Soft skills, such as emotional and behavioral awareness, are far more impactful than technical competency alone. This is what separates the average from the top performers."

Working with Recruiters

There are two main working structures in the relationship between consultants and recruiters. With one, you have your own company that does business with theirs. Both companies interact 'entity to entity'. With the other, you work as an employee of the recruiting company and serve the client on its behalf. Let's take a further look at both options.

Corp-to-Corp versus W-2

A corporation-to-corporation relationship (corp-to-corp) exists when consultants or contractors wish to be paid their entire bill rate without taxes being withheld. This means that they are taking responsibility to deduct and pay their own taxes independently. Many independent workers don't choose this route, preferring to have their taxes withheld by the company and receive a W-2 at the end of the year. Although the W-2 route is more straightforward and easy to maintain, there are some

> **Résumöre®**
> The upgrade to your résumé that highlights who you are and what makes you a good fit for the job.
> www.resumore.com

significant benefits to consider when choosing one that works best for you:

W-2 Status
- Appropriate for most contractors
    - If your goal is to work independently for a short time and eventually be hired as a full-time employee, a W-2 is probably your best bet.
- Appropriate taxes automatically deducted and paid to the IRS
    - If you do not have a company and have no desire to maintain payroll, separating your personal and business expenses indefinitely, a W-2 working status may suit you best.
- Ability to do your taxes the same way as always, once per year
    - If you are not willing to maintain quarterly tax payments and keep a close eye on profit versus expenses, a W-2 relationship may be appropriate for you.

Corp-to-Corp Status
- Appropriate for owners of small businesses and those who wish to build their career as independent consultants

- If you have a desire to truly be your own boss and establish your services as an independent entity, this may be for you.

➢ Can take advantage of the benefits of being a separate organization
- Organizations can apply for an independent tax identification number, which allows you and your company to become interdependent.
- Expenses can be paid at the company level and deducted from earnings, lowering your taxable income.

➢ Will ultimately put more money in your pocket than with a W-2 working relationship
- It requires more work but many services can be outsourced, resulting in costs that are then tax-deductible.

I have this crazy idea that everyone can become successful at whatever they do as long as they are willing to learn, unlearn, and relearn. People have to stop saying "I'm not good at that." Maybe you weren't good at something a long time ago but that doesn't define who you are today or who you can become tomorrow.

> Agile changed my life by providing a framework for prioritizing and pursuing my ambitions. By embedding a few of Scrum's best practices into my day to day routine, I have enriched my personal life in ways unimaginable.
>
> As stated by renowned leadership expert John C. Maxwell, "Time management is an oxymoron. Time is beyond our control, and the clock keeps ticking regardless of how we lead our lives. Priority management is the answer to maximizing the time we have."

Chapter 4

Agile Frameworks for Achieving Your Personal and Professional Goals

Scrumming Your Way to Success in Everything You Do

> *To* succeed in life you need two things:
> Ignorance and confidence.
>
> —Mark Twain

D. Ray Freeman

Rafal Olechowski/shutterstock.com

Building Your Backlog Is a Continuous Process

How do you get started building your own backlog? The first step is as simple as **ABC**, which in this case stands for **Active Backlog Conditioning**. In a corporate setting, separate projects have separate backlogs, managed independently or subordinate to one another. For Agile to be acceptable and usable at home, the management process should be simplified. Everything can be maintained as one project, requiring only one backlog. With active backlog conditioning, you continually add and subtract items as they gain and lose importance or value toward your goals. This frequent *grooming* of the backlog will be covered in more detail as we progress.

Along the way, you will discover some Agile terminology as used in its native environment. You will also find some simple, yet interactive ways to bring Agile into your own environment. By integrating one or more of Agile's best practices into your daily routine, you can put your life and career on the fast track to success.

> The **Backlog** is the ever-evolving list of product requirements prioritized by the customer's representative, also known as the product owner. The backlog tells an Agile team which features to implement first. Agile projects typically split the backlog into levels.
>
> - Product backlog (everything)
> - Release backlog (everything that will be released together)
> - Iteration or sprint backlog (everything that will be done towards the upcoming release)
>
> At either level, the backlog is basically a list of features or requirements expressed in terms of user stories. The product owner maintains and prioritizes the items in the backlog and assigns business value accordingly.

Let's Make A Home Scrum Board!

Agile is more kindergarten than college. With this in mind, you should always start out simple. The home scrum board is a visible representation of goals you wish to achieve and activities that are vital to achieving them.

The scrum board can be as simple as a couple of pieces of construction paper attached to a wall or a large corkboard on an easel. Your scrum board should be something that you and your home team members can gather around. It should be located where it can be interacted with daily.

One of the best ways I've found to get others to participate is to make sure they are involved with the creation of the scrum board and have input on all of the activities that go onto it. All of the goals, stories, and tasks should be added to the board using multicolored notes stuck or pinned to the board. Decorate the borders of your board any way you like, but be sure to leave a big, open space for adding several dozen notes and moving them around. The superstick kinds work best.

Make a title section on the left named "Backlog." Make sure everyone is aware that this is a special place and will serve as everyone's personal goal list. Don't worry if it seems primitive at first. Agile should be introduced gradually to prevent overwhelming those who are new to it.

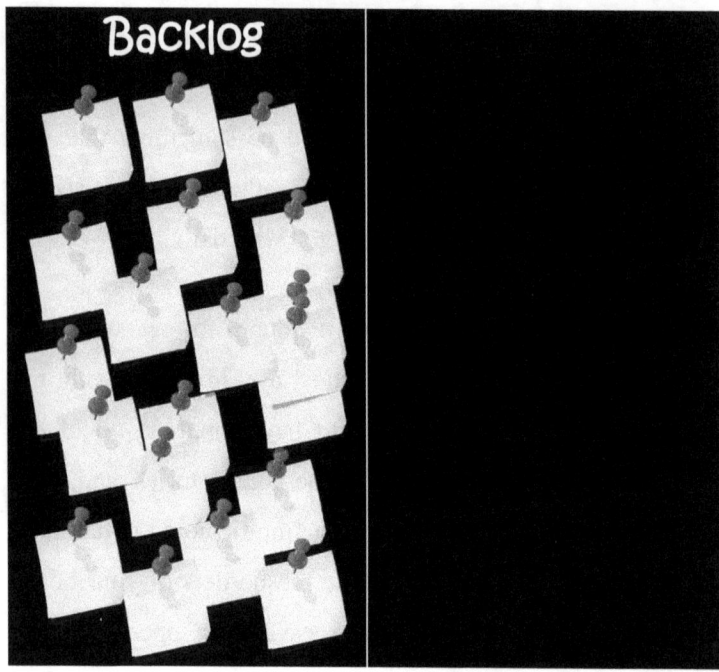

In my house, we tacked two presentation boards side by side to create a big rectangular section on the wall. They're inexpensive and easy to find at most office supply stores. I knew that my wife would never go for anything ugly being put on the walls of our home, so it was part of our project to make something that was not only functional but that looked good too. We knew that some of the items we put on it would be personal to us, so we put it on the wall in a spare bedroom my wife and I use as an office. We made borders with colored tape and wrote "Backlog" at the top with a colored marker.

In the beginning, we wrote a bunch of to-do items, one per sticky note, and randomly placed them within the borders of the backlog section. This helped start our first ABC session and opened the floor for some collaborative brainstorming.

Examples of a Home Backlog

What kind of stuff should be in your backlog? Each team member should write his or her own notes and attach them to the board. It's important for everyone to actively participate. Be patient as this may take some time to get used to. Here are some typical things that might be added to your backlog in the beginning stages of Agile adoption:

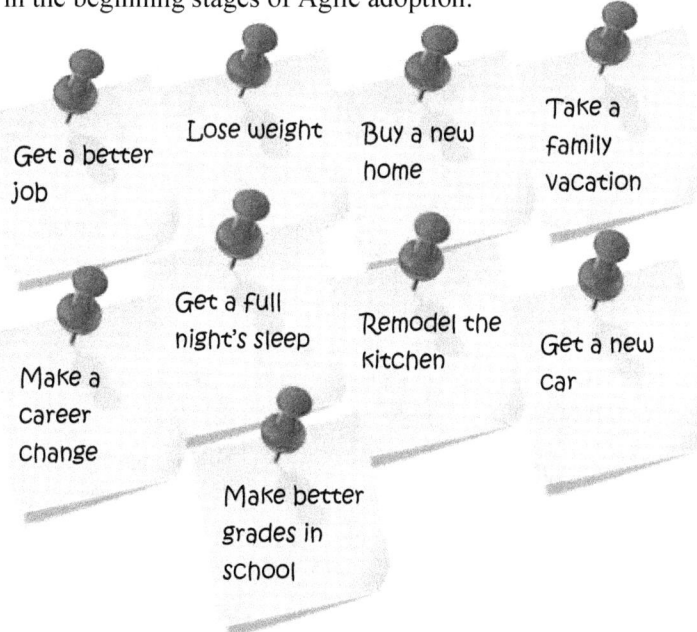

With ABC, this should become a regular and routine activity. Whenever you or someone on your home team comes up with an idea—no matter what it is—your response should be, "That sounds like a great idea! Will you add it to the backlog so we don't forget it?" Most of the time, people try to keep all of their personal priorities in their heads, figuring that nothing will be forgotten or lost, but this is rarely the case. We're human. We forget.

The ABC process requires some level of trust because it can also uncover more personal things occurring in your life such as:

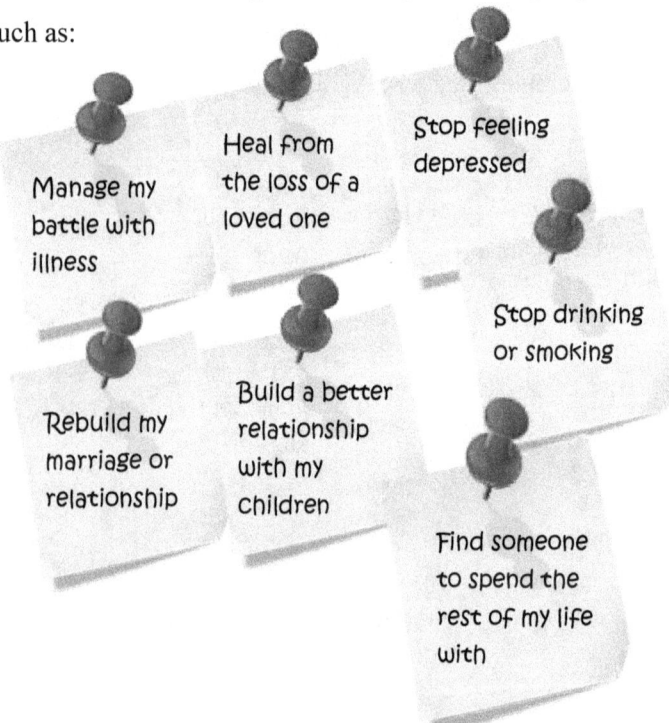

When you first start building your backlog, it may also contain tasks that you already know about but want to make sure you don't forget, like:

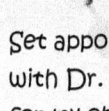

Set up new insurance during open enrollment

Set appointment with Dr. Thomas for my checkup

Fix the hinge on the back fence so the dog won't get out

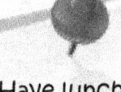

Get school supplies

Have lunch with Tiffany so we can catch up

Some home backlog items may directly relate to your career:

Obtain a new skill to boost my resume

Update my résumé

Land my dream job

Prepare for my interview with XYZ company

Start my own business

User Story In Agile, items in the backlog are called user stories. A user story, often just called a story, is an Agile requirement stated as a sentence or two in plain English. A user story is expressed from the user's point of view and describes a unit of desired functionality.

A user story is not intended to describe every aspect of a feature; instead it is meant to serve as a placeholder for conversation between the product owner and the team. A user story may be written in any form that works for the team; however, the most popular format is this:

As a [type of user,] I want to [do something] so that I [benefit].

Example: "As an **author**, I want to **publish a book** that will **inspire people to improve their lives by being Agile**."

Each story has a set of acceptance criteria that when met constitute completion of the story and its ability to be accepted as "*done*-done." A really big user story is called an "epic", which can be broken down into smaller chunks. To simplify Agile at home, it is not always necessary to write your user stories in such a lengthy format. Don't worry, you won't be graded for accuracy!

Breaking Down the Epics

In the seventh chapter of the seventh book of Matthew, written in an expanded translation of the Christian Bible, there is a popular and empowering scripture. It reads "Keep on asking, and you will receive what you ask for, keep on seeking, and you will find, keep on knocking, and the door will be opened to you." This is an expansion of the older version which simply states "Ask, and it shall be given you; seek, and ye shall find; knock, and it shall be opened unto you". I specifically highlight the expanded version because it emphasises that these processes are repetitions, rather than singular queries with an immediate result. The only way to break down big items is to keep asking questions, keep

> **Backlog Grooming** refers to the process of:
>
> - Adding new user stories to the backlog
> - Prioritizing existing user stories
> - Estimating the size of user stories
> - Breaking down large user stories (epics) into smaller stories or tasks
>
> Backlog grooming is both an ongoing process and the name for a meeting or ceremony itself.

seeking what you're looking, and keep knocking on doors. A user story that contains many unanswered questions is called an epic. It's too big to be tackled on its own. Break it apart by extracting the questions from it. Keep asking questions about the epic until the answer is "I don't know."

When you get to this point and the answers are unclear, unknown or outdated, switch your thinking from "ask mode" to "seek mode". Write down an action item to go find the answer and add it to the backlog. Be sure to write it in the form of an action. Let's examine the epic below which has lots of unanswered questions in it. Then, let's dismantle it into smaller chunks. Along the way, we might even find some new doors to go knock on.

...I want to have a great job so that I can love what I do, make plenty of money and build a meaningful career

Questions to ask:
- What is a great job?
    - Income related
        - How much is the salary?
        - Are there bonuses?
    - Flexibility
        - Is it close to home/school?
        - Can I work from home or travel?
        - Will I be able to take care of my home, kids and personal responsibilities at the same time?
    - Opportunity
        - Will I learn a new skill or meet new people?
        - Is there room for growth or longevity?
    - Meaning
        - At the end of the day, will I feel good about the work I've done
        - Does this have job have a positive impact on the community or the world in any way?
        - Can I imagine myself doing this for a long time?

This list could go on and on but don't get too caught up in the analysis. Write down a few pieces and put them on the board in the backlog where you can see them every day. Remember to write them as action statements. If there are several people contributing to your backlog, put your name or initials on it.

> Make a list of the things that are important to me in a new job. - "TJ"

Getting Others to Participate

It's good to start the ABC process with each person adding items at his or her leisure. It should become a habit, and the board should be a safe place to put anything that's on your mind. For Agile purists, don't worry if these items are not written in traditional story format. Too many strict rules around the process itself may keep others from participating. When the game isn't fun, nobody wants to play with you.

If you are the person who is initiating the Agile process at home, please be patient. Don't expect it to be immediately adopted or functioning flawlessly in one iteration or release.

Most people have never managed their lives this way before. It's new and it will take some getting used to. Be the facilitator. Ask questions.

Lead by Example.

As I mentioned earlier in this chapter, when you are faced with an immense obstacle or overwhelming challenge the best course of action is to "do something". And if that doesn't work, "adjust and do something else". People are more willing to participate when they see that *you* are making a real effort for something better.

> The **Definition of Done** is the universally agreed-upon criterion for what makes a unit of work complete. Done is the state at which all acceptance criteria have been met. Often referred to as "*done-done,*" it is the point at which a unit of work (user story or task) becomes acceptable and potentially shippable to the end user. The definition of done is generally part of the working agreements agreed upon by an Agile team.

The Benefits of Defining "Done"

Defining the meaning of the word "done" is one of the best things you can do to simplify your life in many ways. Has anyone ever asked you to do something but didn't adequately explain what they wanted done or how they wanted it done? All of us have had this experience at some point in our lives.

Running The Numbers

At work, your boss tells you to "run the numbers on that report". You create a spreadsheet, add up the dollar amounts in each column, insert monthly averages and email him what you believe is the finished product—An Excel spreadsheet. The next day he tells you that it's all wrong and you need to start over. This time, you add up the dollar amounts by category instead of by date. You average them across product lines and email the new spreadsheet to your boss.

Feeling content with your accomplishment, you leave work and head for a local bar to join the happy hour with your colleagues. Just as the bartender slides a frosty mug in front of you, a familiar sound catches your ear. It's your ringtone. Your phone vibrates dangerously close to your untouched glass of sweet libations. The name on the screen—your boss.

It may be five o'clock somewhere but not where you are! Reluctantly, you answer the phone using your best office tone of voice. "Hello how can I help you", you reply. "Where is that report", your boss yells. "I need that PDF file with the bar charts right now but you keep sending me a spreadsheet". "Get back to the office and re-run the numbers right away", he exclaims.

Anytime someone gives you a task to do, always get clarity on what are the acceptance criteria that must be met in order for the task to be considered *done*. This way, you'll know exactly what to do in order to satisfy their request. Assuming will cause you to leave a full glass of your favorite swill stuck to a wet napkin at the bar. Meanwhile, you're headed back to the office to work late, again, running the numbers for the third time.

Imagine how the definition of done can be used with your children. "Clean your room" you say to your teenager, only to come back and see that all of the dirty clothes have been shoved into the closet and the shoes kicked under the bed.

When I was a teenager, my mother worked the night shift starting at six o'clock p.m. and returning at six o'clock a.m.-- sometimes seven day a week! We had an agreement that I would be responsible for cleaning the kitchen before she got home from work in the morning. This was a reasonable request and technically, I had all night to get it done.

One morning Momma came home from what had been a hard night on the job. She was dreadfully exhausted after working fourteen days in a row to keep food on the table for my sister, brother and I. She just wanted to come home and collapse into bed before time to head back for day fifteen.

Six thirty a.m. Mom walks in the garage door next to the kitchen. Although the dishes had been washed, I had left them to dry on the countertop. I had placed the food in the refrigerator but didn't wipe off the stove. I gathered the trash from around the house but left the smelly garbage bags inside next to the kitchen door. The floor wasn't swept and there was a stain from some juice I had wasted the night before. Instead of being greeted by a sweet good morning from my mother, I was abruptly awakened by four very loud, very motivating words—"Get Your Ass Up"!

In my sixteen-year old mind, the task of cleaning the kitchen was done, but apparently it wasn't "*done*-done".

Needless to say, it didn't take me too long to figure out the definition of done. Neither of us wanted this to happen again or for it to become a regular wake up call. So we established a set of working agreements that defined what she really means when she says "clean the kitchen".

We knew nothing of Agile in those days but in hindsight, the application of the concept is clear. Defining done sets expectations between what the current state is and what the future state should be. This could also be expressed in user story format:

Mom's User Story: "As a mother, when I come home from work, I want the kitchen to be cleaned according to my expectations, so I don't get upset and yell at my kids".

Acceptance criteria.
- The dishes must be washed, dried and put back in their respective places.
- The floors must be swept and mopped, especially if something has been spilled.
- The trash must be gathered, bagged and taken outside to the garbage cans.
- The countertops and stove must be wiped clean and all cleaning supplies put away.

If these acceptance criteria are not met, this story can't be considered done. Once this concept penetrated the layers of my thick teenage head, I got it! The criteria had been laid out clearly so there was no excuse for misunderstandings in this area. We had very few problems once expectations were set and understood by both parties.

Reminiscent of the nineteen eighty six comedy, Ferris Bueller's Day Off, I still engaged in my own mischievous youthful activities. In the movie, high school student Mathew Broderick coordinates an elaborate sequence of plans to create one of the most awesome 'skip days' of its time. Before his parents returned from work, 'Ferris' made certain that everything in the house was back in order as expected.

Even if I had thrown an unauthorized party the night before, seasoned with all the flavors of teenage shenanigans, I at least adhered to the basic acceptance criteria of our working agreement—clean up before Momma gets home. Let's call it responsible mischief!

Releasing Your Pathway to Success

A release is a group of items that will be completed within a given timeframe. Releases should have a defined start and end date so you can "time-box" the work. Releases are made from the stories in the backlog that share a common subject matter or theme. For example, if you are having a planning session in December. And one of your goals is to get a new job next year. You can define the "Release One" start date as January 1 and the "Release One" end date as March 31. By doing this, you are committing to yourself and members of your team, that "during this timeframe, I will focus on actions that will lead me to my goal of getting a better job."

You've been breaking down the steps by asking questions and you have a list of action items to do over the next few months. Move the items in the backlog that are career related into the 'Release One' column. The other members of your team should do the same with high priority items they want to complete in the first quarter of next year. There can be multiple themes in a single release especially when several people are involved.

The example below shows a release that has been planned. It has items that multiple team members have decided are of highest value and importance. It has other items that need to be done within that timeframe that you don't want to forget. Primarily, it contains items that appropriately respond to what you identified as your highest priority epic: "I want to have a great job so that I can love what I do, make plenty of money and build a meaningful career".

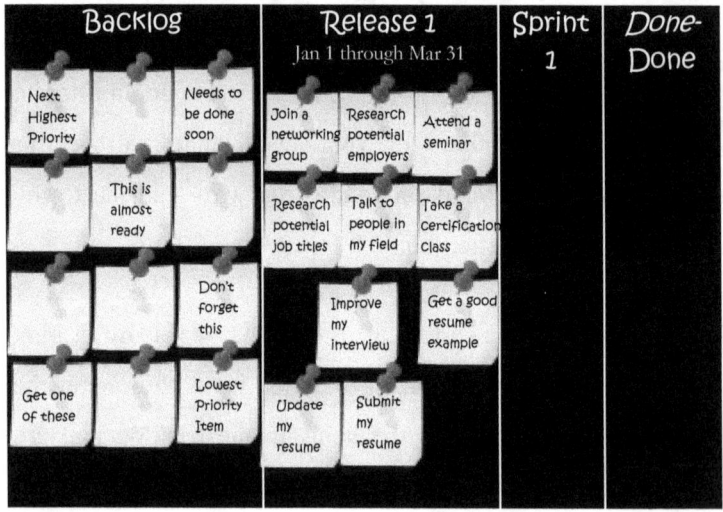

Spikes!

Spikes are the unexpected items that shoot to the top of your priority list. They're usually the result of some new or unexpected occurrence that suddenly became more important than everything else. These things don't usually get noted, but if they were the severity would be obvious.

> Fix the leak in the hot water heater before the whole house floods!

A spike can also be used to unblock an existing story. Sometimes, you can't move forward with one user story because it is being blocked by another. It's OK to say, "I don't know." Therefore, a story can be blocked because you just don't know how to do it yet. As independent workers, my wife and I wanted to establish a tax sheltered savings plan for our business. We'd never done this before and needed to figure out how. Adding a spike was a way to make sure you focus on figuring out how.

> Learn how to establish a 401(k) plan for our company so we can pay fewer taxes and save money for our future.

Get Your Mind Out of the Gutters!

It's Monday morning. As we begin the day, my wife and I have our first scrum of the new iteration. This is day one of the start of something new. Some of the items in the backlog are specific to her, others to me. Items such as "get hair done", obviously don't apply to me but "pickup dry cleaning" could apply to either of us! The most important things are at the top of the list and those that can wait are near the bottom.

> Install gutters around the eaves of the house so the rain water drains properly.

Today, the lowest item at the bottom of the backlog is a note about to installing gutters on the house. This one's been sitting for a while, and unfortunately it's going to be there for a while because I hate getting up on the roof!

It needs to be done but it's not urgent. There are more important things that we could be spending money on right now. Although it's at the bottom of the list it is somewhat important and needs to be addressed within the year. The backlog shouldn't become a stagnant list of broken promises. In this case, I realize that there is a risk to the foundation of our home if runoff is not dispersed properly. As a homeowner, I will protect the vulnerable areas of my home, so I am not struck with the high cost of repair in the future.

As we make more money and have more resources to invest in our home, this note will get moved up higher on the list. Eventually, it will make its way into a release timeframe and be broken down by adding notes like these:

Search the web for a reputable gutter installation company

Setup appointment to get estimates on gutters installation

ABC Gutter installations: 9:00 a.m., 9/24

Add cost of gutter installation to Quicken accounting

The Book Release

One of the stories in my current release is to reach a goal that is time sensitive. It serves as a reminder for me to put extra effort toward a goal that's an important part to my future as a writer.

As an aspiring author, I want to publish my first book by November, 2014.

Although its written on a tiny note, it represents a big chunk of work. Like any big chunk of work, it needs to be broken down into smaller pieces. Before I can publish a book, I obviously have to finish writing it first!

Finish writing the manuscript for my upcoming book

How Do YOU Want to be Communicated With?

As my wife Tahonie and I continue our week, we pretty much work autonomously. We both play an active role in each business, but neither of us likes micromanagement, especially not from each other. Just like any couple, sometimes our differing personalities clash with one another.

A lot of people tell me that "there's no way I could work with my spouse because we have two completely different thinking styles". My reply is that "almost every couple has completely different thinking styles". In fact, most people think that others perceive information the same way they do and a majority of the time, they are wrong.

Recognizing and responding appropriately to personality differences plays a major role in making Agile work for you. Make it a priority to learn ways to better understand how you and your team members think. Don't focus on your individual weaknesses. Identify your strengths and use them collectively to create a stronger whole.

Bringing Work Home

In my line of work, I've attended a lot of training sessions on teambuilding, person to person communications and personality assessments. As the leader of Agile teams for many years, I have experienced the benefits that this type of training brings, even among the most diverse groups of co-workers. Being able to recognize different thinking styles and knowing how to best communicate with them goes a long way in fostering team cohesiveness.

Even if you are not operating a formal business, every family works together on some level. We all have some intersecting responsibilities with the people we share a home with. There came a time when my wife and I needed a little more cohesiveness at home. So we did something about it. And when that didn't work, we adjusted and did something else. Anything worth having is worth fighting for.

The Home Team

A team has two or more people united in pursuit of a common goal. However, a team will never reach its goal if the people on the team are unwilling to work together. One of the working agreements in my home is that we don't settle for average. We are always looking for ways to improve personally and professionally. The following notes were written to address an issue brought up in retrospective. They are items that we both agreed to work on. They represent small, yet positive steps toward the goal of being a happy couple.

> As a couple, we will find better ways to communicate with one another so we can enjoy the time we spend together.

> Find out how we can both take some training classes on how to communication better.

> Find a way for us to understand each other's personality better.

Gaining Stakeholder / Team Buy-In

As the "Agile Evangelist" of your home you will be embarking upon a great adventure. To make the transition from one methodology to another is difficult. To transition from no methodology at all is even more challenging. Share this book and encourage others to participate in the process. Use this as a guide to the basics of Scrum and a framework you can build upon for your own environment. Let this be the catalyst that inspires a conversation—helping you communicate the vision of Agile to your team members.

Whether you are jumping into Agile head first or stealthily implementing some best practices, you must be able to set expectations for those involved in the process. This is a unique way to be Agile because your stakeholders and your team members are the same people. Getting them to actively participate will require some additional effort from everyone. Keep it light, make it fun and reward yourselves often.

The processes described here have been practiced for a multitude of iterations and refined by an equal amount of retrospectives. A guide of this nature did not exist as I began implementing 'Agile@Home'. There was definitely some pushback and confusion at the start. But as we began to see the fruits of our labors take shape, we stuck with it. Don't be afraid to experiment and adapt it to fit your particular environment. Expect your team to go through the four team development stages: 'Forming and storming' in the beginning; 'norming and performing' as your process matures.

Implementing Agile will require some extra effort from those involved. Make a decision that your goals are worth this extra effort and reward yourself when you have completed a significant chunk of it. If you want to live an extraordinary life, you have to do extra things ordinarily. It is no secret that successful people put in more effort than those who are unsuccessful. Successful people value time and effort as the equivalent of currency. Manage your currency wisely by managing it with one of the most effective methodologies in the world.

Reward

People are motivated by the anticipation of receiving recognition for their efforts. Businesses use reward to recognize exceptional effort done by their employees. Reward yourself in similar fashion. Add a story in the release backlog to commemorate it's completion. It can be something as simple as a special celebration dinner with your team members or creating a special certificate to memorialize a significant accomplishment.

Sprint Forward—Two Weeks at a Time

The preceding illustration shows an active scrum board by a mature Agile team. In this example, The Sprint three dates have been added and a blocked story section near the bottom. The items beneath 'blocked 'are stuck for one reason or another and need some attention before they can move forward. These items must be unblocked in Sprint 3 otherwise they will fall into Sprint four. Review them in your next planning session to figure out how to unblock them and move forward.

It is important to note that during the Sprint three timeframe, you should focus only on the items in Sprint three and work them from top to bottom. Don't worry about the items in Release four or the Backlog right now. They're either not urgent, not ready or are simply not going to be worked on right now. Focus on one item at a time in the current sprint moving each note over to *done*-done as it is completed to your agreed upon acceptance criteria.

What Should I do With All Those Notes?

I never would have thought that my life would be managed by a series of little strips of paper! Re-reading some of the older notes reminds me of how far I've come throughout the year. It has become a family tradition to do something special with all of the notes accumulated across releases. For example, all of the notes that went into creating this book will be captured in a bottle and etched with the book's title and date of completion. This represents a significant chunk of work. Not perfect, but done and shippable to end users. The bottled notes will be displayed next to the first published copy of the book in a display case in our living room. If the stories on your notes have played a role in changing your life, memorialize them in a similar fashion.

Retrospectives at Home

Life can be unpredictable. When things go well, we should make an effort to do it that way again. When things don't go well, we should be willing to make an adjustment. This reflection process creates new knowledge that makes the next sprint better than the previous one. This review process should be done at the end of every iteration.

Before you bottle up the notes in the done-done column or construct some creative collage display, Re-read them with your team as part of the retrospective. Conduct a very informal review at the end of every two-week sprint. I emphasize informal because having a casual conversation is usually easier than trying to conduct a formal meeting. Your environment may differ so structure your retrospectives accordingly. Use this time to examine the stories you have accomplished and how they progressed from backlog to release, from release to sprint and from sprint to done. Ask yourself, "If I had to do this again starting today, is there anything that I would do differently" This is a team activity so share this new knowledge with everyone.

During retrospective, you will uncover some new action items and discover valuable nuggets that will benefit your next planning session. This may even help you to establish a family knowledge base of best practices. I believe that the modern Agile family will create repositories of wisdom that can be shared socially across generations.

Retrospectives are face-to-face conversations about what went well, what didn't go so well and what things you will adjust as a means of improvement. "I tried something and it didn't work out so well. I will make the following adjustments and try again". Plan for your next sprint to be better than your last. Reprioritize and keep the process moving.

Keep the backlog alive. When there is empty space in the backlog column, it's ready for more conditioning. There are some learning and brainstorming techniques in the next chapter that are great for extracting new ideas and breaking down complex user stories.

Chapter 5

Learning over Education

The DALA Principle: Developing a Learning Aptitude toward Your Goals

Men fear thought as they fear nothing else on earth—more than ruin, more even than death. Thought is subversive and revolutionary, destructive and terrible; thought is merciless to privilege, established institutions, and comfortable habit; thought is anarchic and lawless, indifferent to authority, careless of the well-tried wisdom of the ages.

Thought looks into the pit of hell and is not afraid. Thought is great and swift and free, the light of the world, and the chief glory of man.

—Bertrand Russell

alexmillos/shutterstock.com

DALA's and Sense!

Derived from the lyrics of urban music, comes a popular idiom often used in casual conversation. "If it don't make dollars; It don't make sense." As grammatically disturbing as the phrase may be, it does contain a noteworthy message. The clever wordplay of this phrase gives the impression that the word "sense" is used synonymous with its homonym "cents."

When used lyrically, this phrase implies that engaging in an activity which produces no income is senseless. I have repurposed this double entendre anew. Although still pronounced "dollars", I use the acronym "DALA", which represents the words "Developing A Learning Aptitude." It takes a "DALA" to make sense!

This statement now implies something a little different. Engaging in repeated activities without learning anything new is senseless. I call this the DALA Principle.

Dollars and Change!

The wordplay continues. The U.S. Dollar is the sum of one-hundred cents. That's a lot of change! You can create additional dollars with the accumulation of such change.

"Change" may also be defined as the process of doing something different. Want to create more dollars? Try generating some change. Accumulate it as one would accumulate 'cents'. More change—more dollars—more sense. Become a facilitator of change and you will be able to make sense of just about anything that comes your way.

Teaching Yourself to Change

The absolute best way facilitate change in your life is to "Develop A Learning Aptitude." DALA! Being Agile encourages continuous learning by repetitively seeking answers to questions left unanswered. Own the responsibility of teaching yourself new things as your projects move forward. Improve your ability to learn by dedicating time and applying your efforts toward the improvement of your abilities to learn and communicate more effectively.

> I want to be able to quickly adapt to changes that occur in my life, so I can learn how to handle any adversity that comes my way.

> Find ways to learn and apply new things quickly and inexpensively so I can be prepared for change.

Agile and the DALA Principle

To embrace the DALA principle, you must find ways to turn learning into a habit. Learning must become something that you want to do, like to do, and seek ways to do more often. Agile is a process of continuous improvement which loses all meaning if you are not willing to continuously improve.

It seems to me that people are rarely satisfied with the way they look, feel or how much money or possessions they have, but they are usually content with what they already know. This is a mistake because the world we live in is constantly changing. Advancements in technology, communication, education, nutrition, medicine—year after year, something new replaces the old, rendering it obsolete.

A mobile phone, for example, becomes outdated within a few months and archaic within a few years.

When I first began my career, I would visit a library or a bookstore to get new information, both of which now travel with me in my pocket everywhere I go. I was not a very good student throughout my childhood, and man do I have the report cards to prove it! Memories of this often steered me away from learning opportunities as an adult. I figured that that since it didn't work well for me then, I'd probably experience the same results today. The real turning point was finding the courage to try again, differently, to not let the past determine the future.

Learning is not just for those enrolled in school; it is the currency that establishes a wealth of understanding. Just as the spending power of money is diminished by inflation, thus it is true in regard to knowledge. In almost every profession, there are continuing education requirements so the practitioner can remain relevant in his or her field. The same principle applies to just about everything you do in life.

According to revered psychologist Dr. Robert Plutchik, author of several groundbreaking theories on emotion and experience, "the human brain has the instinctive ability to change based on the repetition of emotions we experience. If you commit to doing something consistently for a period of time, your brain will establish the necessary connections to embrace it as part of your habitual routine." In other words, you can train your brain to build new habits, including the habit of learning.

Don't ever allow yourself to be deluded by negative past experiences with learning. You can 'rewire' your brain by simply giving it some new habits to build upon. With about two iterations of effort, you can perform some simple actions that will transform you into a habitual learner.

Let's look at some ways that you can develop or enhance your habit of learning. Whether you are currently enrolled in an instructor-lead program or if you are just looking to brush up some new skills on your own. If you have made it to this page in the book you're reading now, can develop a learning aptitude and can encourage others to do the same.

Distinguish Learning from Education or Credentials

I often hear adults say that they plan to go back to school to get a degree. Something may have been lacking in their educational profile. I typically ask, "are you looking to secure a document or do you want to learn something new"? Unenthusiastically, the answers I get are usually more about the document than the acquisition of new knowledge. People often believe that possession of a document will help them conquer some work related fear or dissatisfaction such as:

- Lack of pay or benefits
- Fear of being laid off or loss of employment
- Escape from an unsatisfying career

Oftentimes, this "I want to go back to school" speech is repeated over and over for years. They start and they stop and sometimes start again. But something always gets in the way. The kids, the job, the bills, time, money, ex-wives and ex-husbands... Every time they make a plan, they already have the perfect excuse why it can't come true. We all have responsibilities. Sometimes you have to get creative.

"What are you doing about it now?" If it's knowledge that you seek, there is no shortage of learning opportunities available right at your fingertips daily. It's acceptable to crawl before walking. If the process of obtaining education is cost or time prohibitive, find another way. Don't wait for the perfect opportunity to start a lengthy, expensive process of education. Start learning today. The ambition and drive of the autonomous learner is more desirable to employers than that of the credential seeking sloth.

People tend to believe that the results of an education will provide a golden ticket to financial paradise. They think that opportunities will come find them once they have obtained the necessary documentation. This is a big mistake in thinking. Although formal education is valuable and having credentials will unlock many doors, it will not open them for you nor thrust you through them. The traditional route of education pursued or desired by some, provides only a static solution in a dynamic world. Change is inevitable. Without a learning aptitude, you may be seeking an outdated solution that doesn't address your concerns today. Never stop learning. Continuous mental stimulation lowers the risk of cognitive decline disorders such as Alzheimer's. Use it now or you may lose it later. Plus, the cost of education can be very expensive. Learning only costs a DALA!

Education is an Epic; Learning is a Task

You don't have to join an expensive, complicated diet program to learn how to lose weight or be healthier. You don't have to pursue a degree in psychology to be able to engage your mind, thinking preferences, or behavioral traits. As long as you have questions, there are a number of resources available to will help you find answers. You just have to be willing to look.

Join the Book a Month Club

Pick a topic, any topic. Read one book per month on that topic. In twelve months, you will have read a dozen books on that topic. In one year, you will have more knowledge on that topic than over ninety percent of the population. This new knowledge can open the door to new personal and professional opportunities that you never would have discovered before reading.

Personality and Thinking-Style Assessment

You can more easily develop a learning aptitude when you have some understanding of how your learning style and behavioral styles interact with one another. The better you know yourself, the better equipped you are to engage and challenge your own mind. There are dozens of tools, that can help you absorb information in a way that is most relevant to you as an individual. Personality assessment are frequently offered to employees as part of their team-building initiatives, but this type of learning is also beneficial on a personal level and should be considered vital for the independent worker.

If this type of learning opportunity is not offered at your current place of employment, chapter 10, Resources Recommended by the Author, has more information on products and services of this nature. To learn more visit www.AgileConsultantGuide.com.

By getting to know your brain more intimately, You will be able to tap into my strengths more powerfully. Although you cannot extinguish all weaknesses, having this understanding will provide you with the tools necessary to gain better control of your weaknesses and exchange them for additional strengths.

Online Learning, Open Courseware

I'm not a fan of online university degrees; however, I'm a big proponent of self-directed learning via the Internet. I'm always amazed by just how much is available to the average person with a computer and an Internet connection. I'm equally amazed by how much learning is available for little or no cost.

Did you know that many universities have entire curricula available to the public for free? Prestigious institutions such as Yale, MIT, Harvard, and Oxford offer "open courseware" classes available to anyone online. The Open Education Consortium (www.oeconsortium.org) has classes that are available at no cost to anyone who wants to learn. Many of them have recordings of actual class lectures and in-class discussions available for download, including the course reading materials and exercises.

Open education is a driver for lifelong learning. It offers a flexible education solution that is available on demand. You can give your life and career an intellectual boost whenever you want to.

Over the years of studying team dynamics, I have become fascinated by the inner workings of the human brain. I enjoy learning about how the brain functions physically and psychologically. Beyond reading dozens of books and exploring the web for the latest advancements in intellectual research, I often listen to recorded lectures on the brain and human behavior.

Although this method of learning will never achieve the academic proficiency earned by a university education, the familiarity itself is beneficial both personally and professionally. Gaining this knowledge not only satisfies my thirst for data, it also expands my ability to connect with people on various levels and motivate them more effectively. It helps me perform my job better and helps me to better understand myself and others.

Audio-Learning Integration

I have to admit, this stuff is really boring! Who wants to sit and listen to some guy talking for an hour? Enduring the tedium of an audio lecture on neuroplasticity is not my idea of fun entertainment. I'd much rather be watching football and having a beer! When I first started taking open education classes, I really couldn't focus on them. My mind kept wandering to other things. I had to find creative ways to make it part of my ongoing learning objectives. While reading or listening to such complex information, I found it much easier to grasp if I played instrumental music at low volume in the background.

Even this became a topic I was curious to learn more about. During my research, I discovered that there are many studies supporting the efficacy of beats, rhythm, melody, harmony, tempo and timbre on the learning abilities of the human mind. The book "Take Two CD's and Call Me In the Morning" by Suzanne E. Jonas is a great resource for learning more about music and its effect on the mind.

The most notable scientific research conclusions involve classical selections with higher pitched frequencies and sixty beats per minute tempos. Several movements by Baroque and Mozart fit well in this category. I enjoy classical music from time to time, but most of my playlists are filled with hip hop, reggae, and straight-ahead jazz.

While I'm in what I refer to as "learning mode", I like to experiment with other types of music beyond those that have received the benefit of advanced study. New, hip music that keeps me interested but not distracted. I've found that this enhances my ability to focus on the material for longer periods of time. I seem to retain more of what I learned and am more creative when writing or brainstorming as well.

While researching companies that use Agile most successfully, I found a great example in a company named "Spotify". As a company, they convey a robust and mature Agile culture which is widely respected in the Agile community, they also offer a music streaming solution that delivers the right music for every moment. As a listener, you can customize stations that will adapt to your preferences.

To better fit my personality and learning style, I created stations based on free-thinking hip-hop artist Denis "Gramatik" Jasarevic and the down tempo sounds by artist Simon "Bonobo" Green. Both offer what many refer to as a "chill out" experience in their music. It's entertaining but not overly distracting. Many of their songs contain very few lyrics and are usually at a slower, more relaxed pace. The rhythms of chill out music and be quite stimulating to the active mind.

This combination of styles creates instrumental masterpieces for those who enjoy the syncopation of hard drum beats, melodic bass lines, live instruments and electronic frequencies. Spotify makes it easy to build and maintain the audio ambiance complementary to your needs. Create a task to discover what type of background music works best for you and start building your stations today. Or, you can check out some of mine through the social media feeds of my website www.agilechangedmylife.com.

Time, Perseverance and Recognition

Developing a learning aptitude takes a little bit of time, a willingness to persevere, and the ability to recognize the value of what you'll get in return. Agile provides a great framework to manage and mature all three.

- A little bit of time
    - When you take an iterative approach, you are only focusing on what you will accomplish each day toward a piece of your goal that can be completed within a two-week period.
- Willingness to persevere

- When you make a commitment to do something and find that you are unable to do it, you will recognize that this is stuck behind a roadblock. Until that roadblock is removed, you will not progress beyond the point you are at today.
- When you stop and reflect on what was completed in the previous iteration, you will discover things you didn't know before, or perhaps you will have a new understanding of them.
- Allow this to be your driver and motivator toward your goals because the more you discover, the more light you shed on your path.

- Value what you will get in return
  - Keep your eyes on the prize. In Agile, the main goal is to achieve the maximum value by doing things that add the highest value first. Items of little value will fall backward in life's backlog of priorities.
  - Be sure your epics contain your "why." What do you really want out of this, and why are you doing it at all?

Thinking and Capturing Thoughts

Another way you can develop a learning aptitude is by setting aside some time that is dedicated just to thinking and capturing your thoughts. Get a huge notepad you can attach to the wall and few colored Sharpie pens. Make sure it's easily accessible and that the pens are always close by. Get into the habit of scribing ideas as they come to your head. Always keep sticky notes handy for jotting down ideas, epics, stories, and tasks that come to mind at random. Taking notes on your phone, tablet, or computer is good too, but there is a neural connection that occurs in the brain when you physically write things down with your dominant hand and periodically review what you wrote.

Mind mapping and clustering are excellent ways to capture ideas, complementary to the brain's natural characteristics. Although they may sound sophisticated, they're both very simple to do. Most people think that our brains store information like a video camera, able to replay sight and sound exactly as it was originally experienced. This is not true. The human brain actually collects and categorizes information in clusters, connected by branches.

Mind Mapping to Produce Epics, Stories, and Tasks

Write down a topic that you want to brainstorm or elaborate upon in the center of a large notepad and draw a box around it. Don't worry about perfection or choosing the most perfect word or topic. The idea is to get ideas out. Outside the box, write down the first words that come to mind regarding that topic. Draw a box around it and a line connecting its related box.

Continue to add ideas and expand this image through subdivisions like you'd see in an organizational flowchart. You can draw pictures and icons or use various pen colors to represent ideas and information. Just keep on expanding. Make your brainstorming event appropriate to your audience. It should be a lighthearted event, but make sure that these three areas are covered at a minimum: health, wealth, and well-being. If you're more technically savvy or want to share your map with others digitally, consider a mind mapping software program. One of my favorites is Xmind, which has a free version available for download at www.xmind.net. The image below is low-tech version showing how a central idea can be expanded upon to break it up into smaller pieces.

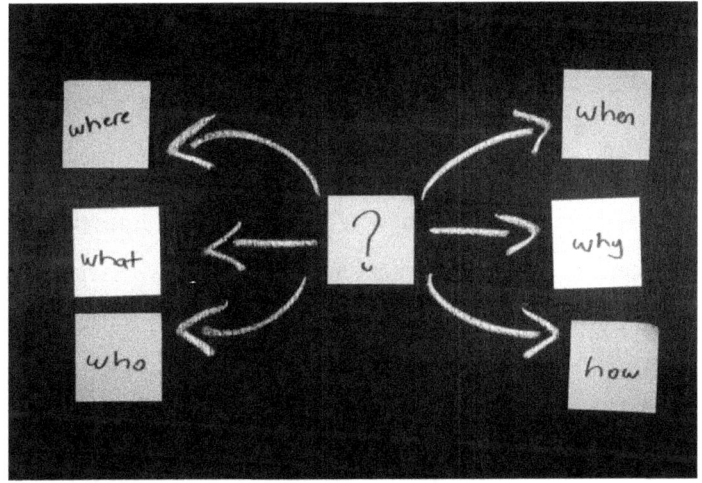

Mert Toker/shutterstock.com

Clustering

Clustering is a great collaborative tool and an excellent way to capture the ideas of a group. Instead of using shapes, write your ideas on sticky notes. One by one, each person goes to the board and adds a note. Those containing similarities should be grouped in a cluster. Once you change to a new idea, add notes to an empty space on the board and begin a new cluster. The board will look messy with notes all bunched together at random, but this is OK. You can start by focusing on the largest cluster first. Select the highest-value items, starting with the single most urgent one first. Order the rest of them accordingly, then move on to the next largest cluster and do the same.

Read! One Way or Another

You can quickly develop a learning aptitude by reading. If you are not an avid reader, remember that there are no wrong ways to read a book. You don't have to start at the beginning and read sequentially to the end. Flip through chapters in whatever order interests you most. Ask other people what they're reading. Spark your interests, especially regarding things that are related to your goals in life. Keep a couple of books saved on your phone. When you have a few free minutes, sit and read a few pages. Don't worry; there won't be a test!

If you spend a significant amount of time in the car, it's a good idea to download a couple of audiobooks to listen to during the ride. Some people learn better by hearing than reading, and if you happen to be stuck in traffic or sitting on a plane, there are thousands of self-help books in audio format to add value to your time.

Depending on the topic of choice, audiobooks can be boring. So consider this a form of training for your brain rather than a form of entertainment. You have to get into the habit of inserting new data into your mind on a regular basis. A majority of the titles available on Amazon are also available in audio format. Some of the narrators are really entertaining in their delivery. I find myself enjoying some audiobooks even more than the printed version. Creating the audio version of this book is in my backlog, destined for a future release in 2015.

Social Media and the Web

The Internet is saturated with information; some is valuable and some is garbage. As much as I love sifting through Internet garbage, I had to change my focus in order to develop a learning aptitude. Social media is good for more than just vacation pictures and funny cat videos! When there are things you wish to accomplish in your life, there are always other people doing or interested in the same. Bypass some of the posts you would typically review in favor of those that are more aligned with your goals.

If your goals are family oriented, join family-oriented groups in your social media circles. Read and share articles, synopses of books, or ideas on parenting, child development, health or education. If you are pursuing career goals, join groups of people already in that career field. Post questions to ask people how they got started in their career. Request an online interview with someone. People are always happy to talk about themselves, and most are willing to help when someone asks.

Watch Wiser TV

Don't get screen poisoning! For many of us, some form of screen has been our entertainer, educator, and confidant throughout most of our lives. I'm referring to the screens of TVs, computers, tablets, phones, and other devices. We humans are easily influenced by what we hear and see played out in front of us. Developing a learning aptitude does not require you to avoid your favorite screens, but you will have to take control of what's on them.

What you watch ultimately changes the way you feel. Just watching the evening news can run you through a maze of different emotions, affecting your sleep, thought patterns, and even the conversation you have with the people in your life. Your brain will accept what you give it. Craft your screen viewing to align with your goals. There is some good stuff on TV but you have to be selective of what you choose to watch. Sometimes the learning channels offer little learning. Discovery channels offer little discovery. History channels are often lean on history. While you are in development of your learning aptitude, consider yourself in training. So limit your exposure to excess advertising from watching too much TV. Yes, you still have to set limits as an adult!

Since the Internet and the TV are so closely related, always dedicate a portion of your TV time to streaming some useful information. If one of your goals or epics involves eating a healthy diet, you should also have a task that says, "Find out what a healthy diet is." In Scrum terms, this means that within this iteration, you will make an effort to learn how to answer the question, "What is a healthy diet?" That can be broken down further to ask what a healthy diet is for me, my individual family members, and even my dog! Depending on how you think, you may even dig deeper into the definition of the word "healthy."

Set a task to find and watch healthy cooking shows. By watching these, you'll learn some tools and techniques to help you prepare healthier dishes. After some repetition, your brain will store and recall these images and information. This activity will increase your vocabulary and visual interpretation of particular foods and cooking techniques that promote the health benefits you seek. You probably won't get the same effect by watching eating competitions or the "Fried Bread, Bacon, Butter, and Sugar Hour!"

Many TV shows are not found on TV. It may take a few Internet searches, but you can easily find several documentaries and presentations on agriculture or the processing of raw ingredients. There's also plenty of programs on foods for disease prevention or stories of triumph by people who have changed their lives by eating differently.

Most of my work is in the field of technology and motivational leadership development. Several years ago, I discovered the TED organization, which broadcasts daily lectures from prominent minds on emerging ideas in technology, entertainment, and design (hence the acronym TED). I added their app to my devices so I could stream a regular dose of information whenever I found a little free time. It helped me find information that was useful toward my goals and enlightened me on things that other people have experienced along their journeys. There are thousands of channels available on YouTube or other online/TV media, including a myriad of how-to videos waiting to be absorbed by your brain.

"Rich people have small TVs and big libraries, and poor people have small libraries and big TVs."
—Zig Ziglar

Chapter 6

From Now On, I Will Plan My Day in an Agile Way

Realizing Your Dreams One Chunk at a Time

Whosoever desires constant success must change his conduct with the times.

—Niccolò Machiavelli

Redwall/shutterstock.com

What do you do every day? Are the things that you do geared toward the success of you, your family, and those around you? Do you feel like the things that you do are moving your life in the direction you want it to go? Are you driving your career, or is your career driving you? Are you in control of your health or at the mercy of chance?

I like talking about success with people. In fact, it's one of my favorite topics. Goals, dreams, hopes, wishes, passions, ambitions, and creative ways of pursuing them are always pleasing to my ears. But sometimes, the conversation turns sour. And it's usually around the same point. After hearing about what you want to do, I always ask, "What are you doing about it now?" And it sucks to hear this as the answer: "Nothing."

What Are You Doing about It Now?

It's a question you should have an answer for every day. You can emphasize certain words to make it more impactful. What are *you* doing about it now? *What* are you doing about it now? What are you doing about it *now*? "Nothing" should be an unacceptable answer from now on. Opportunities will not fall into your hands from the sky. When you step onto a scale, the pounds won't magically decrease without your conscious effort. I hate to burst your bubble, but you probably won't be winning the lotto any time soon either. And guess what? If you hate your job today, working there five more years won't make it any better for you then.

So what are you doing about it now? Every day presents a set of new challenges. Problems may subside temporarily, but a fresh batch of new problems is waiting just around the corner. You need a plan and it should be flexible enough to change based on the most current situation. Revise and share your plan every day for what you want to accomplish that day. Your daily plan should consist of the actions you will take today toward your goals of tomorrow. It's a set of commitments you make to yourself, and it creates greater accountability amongst teams. What did you do yesterday toward your goals? What will you do today?

> **The Daily Scrum:**, also called a stand-up, is a fifteen-minute or shorter update in which members of a team address three key questions:
>
> 1. What did you get done since the last stand-up?
> 2. What will you get done before the next stand-up?
> 3. Are there any roadblocks in the way of what you need to get done?
>
> The meeting is typically held at the same time and in the same place every day, with team members usually standing in a small circle. This stand-up focuses on coordination through collaboration and the meeting serves to raise the visibility of each person's work and to ensure work integration.

Let's Have Scrum Today!

At home, my wife and I have scrum every morning. Our scrums are not as formal as those in work environments; however, we do maintain a similar structure.

Basically, we talk about what we did yesterday, what we're doing today, and anything blocking us from moving forward. We both have busy schedules so this usually occurs just after the first cup of coffee. With a devilish grin, I jokingly ask "Hey babe, how about we have a little scrum this morning?" But my wife is not a morning person so sometimes that joke doesn't fly so well; then I'm left to have scrum by myself!

"Scrum" gets its name from the sport of rugby. It's basically a huddle like you see in American football. It's a quick meeting between team members that happens every time the ball goes out of play. Even though football and rugby teams both go through several hours of practice sessions before any game, they still need to communicate in the moment.

Then what is this scrum, this huddle? Why do they need to stop and chat for a minute before every play? Shouldn't they just stick to the plan? Why are the coaches and coordinators still making crude, hand-drawn strategies in the middle of a game that has already been strategized? It's because they realize that you can't know everything ahead of time. Time brings about change, and the team needs a quick get-together to reassess the current state of reality. Where are we as a result of the last play? What are we going to do in the next play?

Standing at the scrum board, the Freeman Motivation Team, my wife and I, will take a quick look what sticky notes we can work on and complete today. Even though we have a plan, we still need to huddle daily. We don't go as far as dictating each other's every move, but we have a lot of common goals that we can coordinate and help each other with. Your family may be structured differently than mine, but your planning sessions will uncover the daily goals and actions items you have in common.

Happy Brainstorming Day!

Every few weeks, my wife and I have a brainstorming session to update our want list. We strive to have fun with it, so don't think for a minute that these are stuffy business meetings. It's a part of Active Backlog Conditioning we call brainstorming day. We'll make turkey sliders and sweet potato fries and definitely crack a bottle of wine or two. Some upbeat music will play in the background, but TVs and phones are turned off. The puppy will get a nice beef bone to chew on to join the party. It might seem a little silly the first time, but once you've seen the outcome of this exercise you'll welcome brainstorming day as a vital part of your ABC routine.

Agile Trains The Brain

Our brains will make every attempt to craft the reality that we seek. But it won't happen by osmosis. You have to do stuff! You have to say what you want out loud, write it down, and talk about it. Dreams come closer to becoming real that way. "Time-boxing" helps to make them arrive quicker. When you plan your release, you have created a time-box. Your brain will align itself with the commitments you make to yourself and others. The brain is malleable and will mold itself to habitually seek the input you give it. Along the course of a release timeframe, your brain will be learning new habits and constructing creative pathways to your success.

Seek success in everything you do. Be accountable to yourself as well as to your team. The daily scrum is a recurring commitment to your "whys." The "whys" are the reasons you are going through this extra effort. They also give you the motivation to keep moving forward. Sounds like a lot of work but there is a good reason for doing it.

Finding Your "Why"

Why go through all this trouble? Daily meetings, planning sessions, brainstorming, backlog conditioning… Is it really worth it? You tell me. When most people are queried about why they do something, the word "want" is usually part of the answer. Quickly write down fifteen things that you want. They can be tangible or intangible, long or short term, and personal or for someone else. Here's an example, but you should definitely write your own "I want" list in an effort to find your "why":

I want:
- Better communication at home
- More patience
- More money
- More time to enjoy life
- Better opportunities for myself
- A new house or car
- To feel better
- To help people
- Better opportunities for my children
- To lose weight
- To sleep
- To lessen the suffering of others
- To be successful
- To look or feel more attractive
- To travel more

Are any of these things on this list really worth your time and effort? If so, and if you ever wish to reach these goals, these items will become projects and features of who you are and who you want to become.

The items on your want list will remain out of reach until you are willing to do something about it. When you are pursuing your goals with such dedication, people may think that you "work" too much. And sometimes this will be true. You will work more than the average person but the definition of the word "work" will have new meaning. When I say to someone, "Sorry, I can't join you today; I have to work," I'm not talking about punching a clock at some place I don't want to be. That means I'm putting my efforts toward something on my "want" list.

Work hard, work smart, and always be thinking about the future. But don't forget to play hard too. Personally, my "why" is because I love to travel, I love talking to interesting people, and I really want to have a positive impact on people's lives. Oh, and don't get me wrong—I want money and the houses and cars too! Your hard work should be rewarding to you, and it should give you reason to celebrate and enjoy life to the fullest.

Your planning efforts and the adjustments you make along the path will turn hard work into smart work. Embrace your want list. It's yours. Don't minimize it or meter it in any way. Aim high, and don't be afraid to take the stairs or climb a few ropes to get there.

Communication is my focus. I want to be able to communicate on many levels, in multiple languages. I want to live my life in such a way that I am continuously moving toward these goals. I want to create opportunities that get me closer to those goals and have benefit to me and the people around me in the short term and beyond.

This isn't too much to ask either. Why not plan? Plan as if your life depends on it, because in a sense, it does. Dream big, not small. I'd like a Denver omelet in Denver and a New York pizza in New York. I want to work on my Spanish in Spain, my French in France, and my English in England. On team Freeman Motivation, we have established that we will take time to relax and enjoy some of the fruits of our labors. We'll do what it takes to coordinate those labors to produce bigger fruit. And we will work our asses off to make it a feast instead of a snack.

Explaining Agile to the Non-Technical

Waterfall methodology is what gave life to Agile. In fact, it's part of my roots, even before my career ever started. I didn't realize it till much later in life when trying to explain what I did for a living to my seventy-year-old mother. For years when people would ask her what I did for a living, she'd give a standard reply: "Something with computers!"

One day she asked me to explain it to her in detail. "Pug"—this was my childhood nickname—"what exactly do you do for a living?"

I went into my best corporate explanation of Agile methodology blah, blah and could see her eyes starting to glaze over. She nodded often, but I could tell we were getting nowhere. I was trying to tell her how complicated it was but couldn't find the right words to say it. Mom is long retired now, but her memory of the business world stems from her twenty-five years of experience working in a manufacturing environment. As a quality assurance analyst in an aluminum can plant, my momma would be familiar with being part of a process that creates a usable end result; I knew this.

As I pondered for the right words to say, it reminded me of the time she visited my school when I was a kid. It was one of those career day events where parents come in to talk to the kids about what they do for a living. Mom showed up prepared with physical examples everyone could interact with. She laid out various pieces of aluminum in the order they go through along the path to become a can of soda. Actually, she worked for a beer-brewing company, but she made an adjustment for us kids.

I remember her being really good at explaining an extremely complex process in very simple terms, even for a child to understand. She gathered the class into a big circle around the room, giving each one of us a shiny piece of metal in various shapes and forms. She explained how all of these pieces were cut, shaped, cleaned, painted, and eventually filled with your favorite beverage and shipped to the stores. All the kids passed around the parts, and Mom had us to act out the sounds and actions that each machine would make throughout each stage before completion. She talked about the people and the machines that all worked together to make this not only possible, but also able to be done over and over really fast.

I may not have understood it then, but Mom spent her career working on a project not so different from the software projects in my career. She was part of a team that was engaged in a systematic approach to consistently create high-quality results. This symphony of processes is synonymous with those that gave birth to Agile.

When I reminded her of this and associated it with a process she was already familiar with, it was easy to compare it to Agile. I said to her, "Mom, Agile is the process of delivering small pieces of completed work. My job is to lead and support the team throughout the process of delivering those pieces of work. At some point, those pieces fit together to make a new product or service." I could tell that the light bulb had gone on as she started to ask some really good questions. I explained that the difference in my environment is that the people, the process, and even the end products are adaptable as needs change.

"Oh, I see," she replied in an excited voice. "So you could start out with flat pieces of metal and later decide that eight-ounce cans are more desirable than twelve-ounce cans. You wouldn't have to go back, change all the specs, retool the machines, modify the processes, and retrain everybody. You can make changes and keep on moving."

"Exactly," I replied with enthusiasm. "There's a lot of coordination and teamwork that has to exist for this to happen. A big part of my job is building up that teamwork."

Lessons Learned

What I learned from this experience is that any time a new system or process is being presented to someone, it needs to be explained to them in a way that can be associated with something they're already familiar with. In an environment where others are not familiar with it, Agile, much like anything else, must be introduced little by little. Its values and principles can be adopted more naturally by introducing it gradually versus dumping it all on a person at one time.

The people involved must have some common objective to build upon. Often, finding this common objective can be difficult, especially when communications have been damaged. Throughout my quest for further Agility, I discovered some really simple strategies that help to resolve conflict, improve communication, and build strong teams. There would come a time when I would need this at home just as much as at work.

Chapter 7

The Working Agreements of Love

Change Your Routine, Change Your Environment, Change Your Life

> *Coming together is a beginning.*
>
> *Keeping together is progress.*
>
> *Working together is success.*
>
> —Henry Ford

Ferdiperdozniy/shutterstock.com

Can Being Agile Save Your Marriage or Relationship?

I was hesitant about how to approach this topic. I wanted to be transparent and present an adequate depiction of my home life in a way that was descriptive enough to fully engage you as the reader. But at the same time, maintain a sense of privacy in the non-public aspects of my marriage.

Agile, on its own, is not prescription for fixing a broken relationship. However, I was experiencing firsthand how it could bring people closer together—even complete strangers. Surely, it could have a positive effect on people who love each other.

Over the years, I've worked with several different teams, all motivated to achieve a common goal. By applying the principles of Agile, we learned and grew stronger together. We built friendships that have crossed departments, crossed companies, and even crossed continents. You will inevitably spend more time with your coworkers than you will with your spouse, significant other, or family. Once you factor in the commute, preparation for work, and sleep, there is very little time left for the ones you love.

Anyone in any relationship undoubtedly realizes that problems, conflicts, and misunderstandings are inevitable. There is no perfect relationship. They all require work, dedication, and commitment in order to be successful. All relationships can benefit from teambuilding and Agile has just the right principles to help people adjust and refocus on one another on a regular basis.

This is difficult to achieve, even with trained professionals stuck in a room together, doing a job for five days a week. Don't underestimate the complexity of achieving agility in your own relationship, especially if you have both reached the boiling point. But when traditional solutions such as couples counseling had little effect, I knew that we had to start thinking outside of the box.

It was evident that our relationship was in jeopardy and we had reached the realization that if we didn't do something soon, what we had built together would soon fall apart.

When something is worth fighting for, anything can be achieved. There came a point when we would experience this firsthand. I felt like it was her fault, and she felt like it was mine. We both were right—half the time. I remember when my wife and I would arrive home from work in the evening and sit at the kitchen table, swapping stories about the day. We would laugh and chat like best friends who hadn't seen each other in a long time.

But what was once an environment of laughter and expression had quickly become silent. Perhaps it was some unresolved disagreement, pressures at work or maybe just the monotony of doing the same things over and over. Whatever it was, we were definitely giving each other the silent treatment along with a little passive aggressiveness.

Days of this behavior turned to weeks as the fog of discontent grew thicker between us. I knew how to handle this predicament in the office but was clueless of how to deal with it at home.

As the scrum master of Agile teams, whenever communication was at a lapse, we'd change something. Something as simple as a change of environment was sufficient to elevate the mood and get disgruntled co-workers communicating again. I decided to try this at home, sticking to my motto—doing something is better than doing nothing.

One particular evening, a slight modification of environment would allow us to begin the mending process and improve upon it incrementally. It was a Thursday night as we sat down to what was turning out to be another silent evening. The evening news was on TV in the background. The first headline was about a disaster. A fire had claimed someone's life, and the victim's family members were expressing their pain and sorrow. It was a terrible tragedy, and our hearts went out to the victim and family.

We exchanged words like new neighbors chatting between a fence in the yard—with the petty banter of strangers, formal and brief. The TV was doing most of the talking. The next three news headlines were equally depressing. Local stories involving murder, corruption, and theft. The intent of having the television on during dinner was to put an audible blanket over our lapse in communication, but the content was contributing to our discomfort.

Good Mood TV

Something had to change. I looked over at my wife and asked, "Do you really want to watch this right now?" "Not really; put on something funny instead," she replied. Since we hadn't been talking, I hadn't realized that she was equally disturbed by the news content and would've made the same suggestion a long time ago if we had been communicating. We always watch the news but maybe it was time to give it a break for a while.

Those few interchanges lead to a major breakthrough. We agreed; something we hadn't done in a very long time. "We won't watch the news during dinner nor just before bed."

We didn't want the silent treatments anymore either. So we kept the TV on at dinner but played some old comedy shows instead. Maybe if we were laughing at something we'd accidentally stumble our way into a conversation or two. We wanted to bring back the smiles back one way or another. This was 'doing something' taking the place of doing nothing.

We'd been watching a 2004 sketch comedy called the Dave Chappelle Show. One of our favorite episodes was on and it actually had us laughing out loud. We found ourselves smiling at each other again. Even after the show was over we still reminisced on the laughs we'd just shared. "Remember when Dave was pretending to be Rick James?" my wife asked. "It's a celebration!" I replied in my best impression of Dave doing Rick.

I'm not even sure what show came on next because by then it didn't matter. We were enjoying each other's company a half hour at a time. We decided to make dinnertime more fun by doing this every time we sat down to eat. Putting on some 'good mood' entertainment during dinner put us in a better mood at one of the most critical times for family conversation.

Although our problems were not miraculously resolved by this, it did provide a positive step in the right direction. We were taking small steps to address our highest priority item. The results went well so we also decided to do more of it. It opened my eyes to the ways Agile can strengthen any relationship. By committing ourselves to this simple yet unconventional change over the next couple of weeks, we noticed a dramatic change. We were no longer conversing like polite visitors as we had become accustomed to. Instead, we were giggling like teenagers at some of the silliest adult humor you can find.

Laughing together provided the catalyst for kick starting real talks between us. I mentioned to her how what were doing was similar to some of the things Agile teams do. Such as relying on face to face communication and establishing a set of agreements between team members. My wife actually like the idea. She said that "Working agreements are kind of like ground rules." While we were in an agreeable mood, we adopted a couple of new agreements that we both found most important—always greet each other with enthusiasm and never go to bed angry. In fact, we found a wall piece to hang in our bedroom that read, "Always Kiss Me Goodnight." Later, we added another piece of artwork to our kitchen in view of the table where we eat. It read, "Live, Love, and Laugh Often."

Family Time is Our Time

It's hard to remember what you were mad about when you're doubled over in laughter. My wife and I chose to limit the amount of negativity that flowed into our home and increase the positivity that flowed out of it. We decided to never let anyone or anything take your joy away. Even if it means missing the most urgent phone call or text. Simply looking at the phone to see who's calling during dinner can sometimes ruin the whole experience. If you ever call or text one of us during family time, you will not get a response.

> Agile teams adopt working agreements when working together on a project. These agreements define a simple set of ground rules that were created by the team for the benefit of the team and that meet the agreed quality expectations. The working agreements are posted where they can be seen or accessed easily by everyone on the team.

The Positive Environment

Along with the new décor, we decided to paint the walls a different color and make a few upgrades to our living space. We put up a digital picture frame that constantly flips though photos from the trips we've taken. We wanted our environment to glow with one good time after another. Occasionally, we'd walk down the hallway together and stop to gaze at a picture of the Eiffel Tower for a few seconds. In that moment, I can almost smell the scent of fruit filled crêpe's wafting from the street vendors beneath it. It reminds me of the hours we spent basking in its shadow on that beautiful day we were in Paris—July fourth, two thousand and twelve.

In many ways, Agile principles helped us get us on the same page. Not perfect, but something—better than nothing.

We recognized a potentially toxic situation and addressed it before irreparable damage could occur.

Don't ever wait for a situation to improve on its own. It won't. The first step in working together is talking. When you find yourself in a lapse of communication at home, do something that will inspire more conversation. Do something that will make you smile and laugh together. Do something that will make you reminisce. Most important of all, do something.

R & T Freeman –Paris, 2012

> The fifth principle of Agile is to "build projects around motivated individuals; give them the environment and support they need and trust them to get the job done."
>
> AgileManifesto.org

Chapter 8

If You Had More ____, What Would You Do with It?

Commitments, Rewards and a Unique Use of Agile

> *To accomplish great things, we must not only act,*
> *but also dream; not only plan,*
> *but also believe.*
>
> —Anatole France

mypokcik/shutterstock.com

> **User Story:**
> As a husband/wife, I want to be the best man/woman I can possibly be—to love, lead, and provide for my family indefinitely.
>
> **Acceptance Criteria:**
> - Family time must be prioritized above work time. Each person in our family will support one another through difficult times both at work and at home.
> - We will find ways to enjoy ourselves doing the things we like to do and do them more often, as part of our daily course of life.

Being Agile is Self-Sustaining Behavior

Agile is a great way to manage your life project because it promotes behaviors that are self-sustaining, making life more enjoyable. There will always be work to do. There will always be times that you have to work longer or work harder than usual. We strive for a sustainable pace but to do so, there has to be room for pause.

Everybody needs a break. Opportunities for relaxation should be given priority in your backlog. They should be part of every release, every iteration—even if it's something really simple like taking a mental health break. And that's nothing more than time set aside for the purpose of doing nothing: "I will take a little time to clear my mind every week, starting this week."

Challenge Yourself and Prepare Your Own Reward

You are the executive of your life and should treat yourself as such—maybe not as lavish as some, but at least like a VIP. You are a very important person. You deserve more, you deserve better, and you deserve everything that you are willing to focus your energy toward.

> There will always be work to do; there will always be challenges both big and small. Agile changed my life by reminding me of how important it is to pause and reflect on the things that make life so enjoyable.

It's not always easy to know where to focus your energy, how much to focus it, or for how long. What follows is a simple tool you can use to narrow things down a bit.

Ask yourself this question:
"If I had more _____, what would I do with it?"

Always reward hard work. It's a basic human need that you shouldn't deprive yourself of. Find a way to do the things you like. It's OK to mix work with play because one won't last long without the other.

"Time" is one of most popular answers to this question. Say it like you mean it—"If I had more time I would travel." You can use Agile practices to make sure you dedicate time to *not* working. That means planning to fund breaks or vacations and removing any potential roadblocks that get in the way of them! If you want to see the world you can, but the world doesn't come cheap.

"If I **make more money**, I will **travel more**, so that I can **see more of what the world has to offer**."

"I will find a way to **make more money**, so that I can **see the world**. "

"Today, I will **spend at least two hours working on my business**, so that I can **make enough profit** to **take a trip to Europe next year.**"

Fill in the blanks to complete these sentences similar to those above but using your own commitments:

1. If I _____, 2. I will _____ 3. so that I can _____.

4. I will find a way to _____ 5. so that I can _____.

6. Today, I will _____ 7. so that I can _____.

Don't forget to have fun. Speak it, write it and do it. Finding new ways to enjoy life should be part of your routine. I talked to some guys recently that made Agile a key part of how they have fun.

Agile Takes the Stage

I always like hearing about creative ways that people are using Agile. But, I'll never forget the way it was described to me by Ed Grannan, principal of Improving Enterprises. Ed leads a team of Agile professionals through some of the most complex technical strategies for their clients. His latest project still uses Agile but is a whole lot less technical. His team got together and formed what is probably the first Agile band in the world. Being a musician myself, I had to know more, so I asked Ed to give me the scoop on AMP.

The Agile Music Project (AMP)

"It all began with a rocky start and we limped along for a couple of months, rarely agreeing on what to play," Ed related. "Then it struck me like a slap in the face—we do Agile every day for our software development projects. Why not do it for our music? The rest of the band agreed, and then we began the task of converting Agile for software projects to Agile music projects."

In May 2013, a few Improving Enterprises employees (Improvers) started having monthly jam sessions. After a couple of months of "forming and storming," they joined an open mic night at a local bar. "We really didn't have a particular set," Ed told me as he shook his head and smiled. "We just got up on stage and did what we'd done in our jam sessions; one of us started a riff, and the rest joined in."

The AMP band eventually started performing covers of well-known songs, and that evolved to some original material as well. The more they played in front of people, the more they seemed to get into them.

The next time they hit the stage, the guys from "Improving" had improved. They set some band working agreements. They agreed to learn more cover songs and work toward a more refined set list. The next order of business was to coin the name Agile Music Project. The band, AMP, was born.

Ed and I were having this conversation over a huge platter of spicy chicken wings and a couple of tall craft beers. He explained to me how incorporating aspects of the

Scrum framework helped to form and grow the band. "We'd start out by brainstorming on what songs to play and from that we created a backlog," said Ed. "We had some concerns of how to ensure quality, especially with varying degrees of skill and limited time to practice the songs. We decided to use Rocksmith," he continued, "which is a great software program for learning to play popular songs on guitar."

The heartbeat of AMP is Brian, an accomplished drummer who keeps the rhythm going and the people moving their feet. Mike, Josh and Bud are shredding guitars while Ed lays down the bass lines. Vocal duties are shared by all. AMP is hitting the stage with more confidence, sounding great and having a blast in the great town of College Station, Texas.

While forming AMP, they all agreed to be co–product owners and "dot vote" to select a set list from the available songs in the program. They also agreed to hold two-week sprints, during which, they would dedicate time to learning the songs in the backlog. At the end of each sprint they perform a demo, which is basically proving that each person can play their part independently and together with the group. After the demo they have a retrospective to identify areas of improvement and help one another to do so. Then, they create an action plan for what they will do differently in the next sprint. With all of the technical details out of the way, now it's time to jam. And so the band plays on.

The AMP really lives up to its name. They conduct a session where songs are broken down into small pieces, learnable within the sprint timeframe. Each piece is sized according to complexity and given "story points" just as a scrum project would. The team then holds each other accountable to learn their part within the sprint timeframe.

It's been less than a year since the formation of AMP, and the Agile process has done them well. The band sounds great, and they're already planning to record some of their music in upcoming releases. They may not be rock stars yet, but what a cool way to show how Agile can turn a fun idea into reality. What began as a casual team room conversation about the love of making music, evolved to standing ovations at the local bar scene. Good times, AMP! Good times. Check out AMP at www.agilemusicproject.com.

Live life to its fullest. Be bold, be courageous, be different. Make the most of what you have and find creative ways to insert enjoyable experiences into your life.

Agile Music Project

Chapter 9

In Closing

Success is going from failure to failure without loss of enthusiasm.

—Winston Churchill

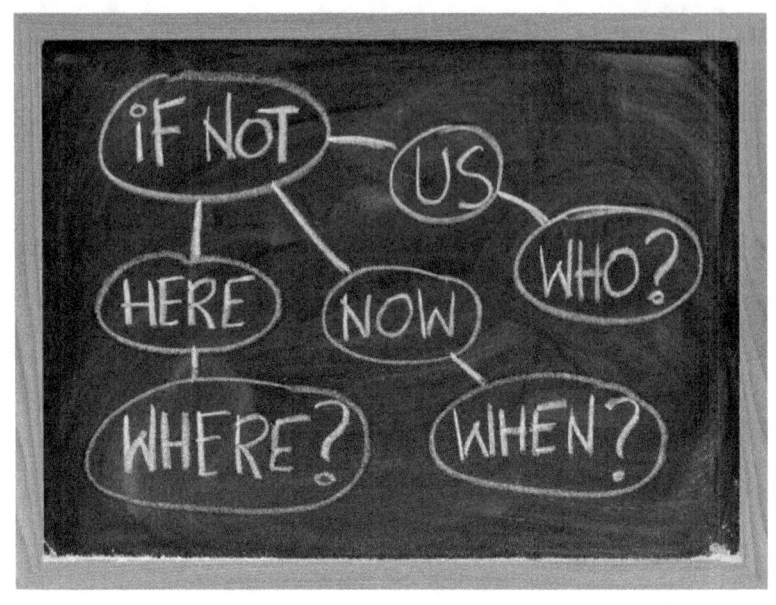
marekuliasz/shutterstock.com

There will come a time when you have to make a decision—one that will ultimately change your life and will determine the future of everything in your life.

If you are ever to change for the better, you must train yourself to take full responsibility for your life's successes, and it's failures. Challenge the way you categorize wants versus needs. The two have a way of bleeding into one another, which is not always a bad thing. There are so many things that influence our thinking, throughout every interaction, during every day of our lives. Guard your mind as if it were a museum. Fill it with priceless artifacts and then open the doors to share with visitors. Leave room to expand, exchange, and upgrade its features. Propel your growth. Set and respect the rules for your mind museum demanding that others do the same.

The difference between those who are great and those who are average is that the great ones are willing to do something that the average will not. That's what makes them great.

Learning methodology is not the key. They key is applying methodology through your strengths, getting better over time, growing, sharing, and experiencing the best of what life has to offer.

Chapter 10

Resources Recommended by the Author

mypokcik/shutterstock.com

Agile Consultant Guide LLC (ACG) Provides the following products and services:

✓ **The Emergenetics Profile**—Pronounced "*emerge-genetics*", the Emergenetics Profile is an instrument that helps:

> ➢ <u>Organizations</u> capitalize on a common language and build a collaborative workforce
> ➢ <u>Individuals</u> gain greater self-awareness and value their strengths
> ➢ <u>Teams</u> learn and understand those whom they work and interact with

It was developed to distinctively measure how people think and behave. With this insight, you can build personal strategies that attain immediate and long lasting results. Emergenetics offers a defined perspective on personality like none other. Developed through years of psychometric research, the Emergenetics Profile accurately measures three behavioral attributes which are:

> ➢ Expressiveness
> ➢ Assertiveness
> ➢ Flexibility

Emergenetics also measure four thinking attributes which are:

> ➢ Analytical
> ➢ Structural
> ➢ Social
> ➢ Conceptual

Contact us at www.acg1.net to get your profile TODAY!

- ✓ **Emergenetics Workshops**—Customized training for organizations, leaders and teams
  - ➢ ACG will partner with you to create a customized workshop or retreat experience for your team to meet your express needs. Our Emergenetics workshops build unparalleled communication skills that result in increased team velocity and improved work and personal relationships for your employees.

- ✓ **Keynote Speaker**—Need a keynote speaker for your next event? Whether 50, 500, 1000 or more – boost the agility of your next meeting with D. Ray Freeman.

Topics include:
- ➢ Being Agile at Home – *Bringing a sense of Agility to your personal life*
- ➢ The New Definition of Hustle – *Taking control of your life by making something out of nothing*
- ➢ A Meeting of the Minds—*Using Emergenetics® to boost team performance*

To book D. Ray Freeman for your next event contact us via email at info@acg1.net.

IT Career Staffing Services

- ✓ **Fortis Talent** is a talent acquisition company that consistently provides its clients with the absolute best strategic assessments and technology-consulting talent services on the planet. Put your best skills to work by contacting them at www.FortisTalent.com.

- ✓ **Connections IT Services** is an IT vendor that superbly aligns strategy with talent. Their professional and personal approach helps to identify, qualify, and retain the best talent for its clients. To benefit from their intimate understanding of corporate strategy and culture, contact them directly at www.connectionsitservices.com.

Learning and Development

- ✓ **Improving Enterprises** provides IT consulting, project outsourcing, training, and networked recruiting to medium and large companies. To create hands on learning experiences in the areas of agile processes, object oriented design and test driven development, contact them at www.improvingenterprises.com.

- ✓ **Open Education Consortium** seeks to scale educational opportunities by taking advantage of the power of the Internet, allowing rapid and essentially free dissemination and enabling people around the world to access knowledge, connect, collaborate, and develop educational approaches that are more responsive to learners' needs. www.OEconsortium.org.

- ✓ For additional independent learning opportunities go to www.ACG1.net and signup for the mailing list to receive up to date content on ways to boost your career.

Professional Services
- ✓ **Ballard Family Dentistry** Your smile is one of the first things that others notice about you. Many people, however, are apprehensive about smiling because of how their teeth look. Ballard can increase your confidence and self-esteem by giving you the smile you've always wanted. www.DoctorBallard.com.

- ✓ **Alliance of Hope for Suicide Survivors** exists for those who have lost a loved one to suicide. They provide healing and compassionate support to those who are suffering through the lonely and tumultuous aftermath of suicide. Their services help people survive and go beyond just surviving, to lead productive lives filled with meaning and joy. For more information or to make a donation please visit them at www.allianceofhope.org

About the Author

For more than eighteen years, D. Ray Freeman has lead international e-commerce and technology programs that span a diverse cross-section of industries such as energy and utilities, transportation and right-of-way development, workforce motivation, and, most recently, tourism and global air travel.

D. Ray Freeman is a professional speaker and behavioral dynamics trainer to Fortune 500 clients internationally. With more than eighteen years of experience in the field of internet technology project management, he instills the principles of Agile to accelerate business and individual effectiveness.

Mr. Freeman resides in Fort Worth, Texas, with his wife of seventeen years, Tahonie, who is an accomplished Realtor for "Honey Realty" serving the Dallas Fort Worth area and their Frenchie Pug puppy—Spency.

Agile Changed My Life

www.ingramcontent.com/pod-product-compliance
Lightning Source LLC
LaVergne TN
LVHW051832080426
835512LV00018B/2839